PCB Trace and Via Currents and Temperatures: The Complete Analysis

* 2nd Edition *

Douglas G Brooks, PhD

With

Dr. Johannes Adam

DEDICATION

This book is dedicated to Dr. Johannes Adam, Leimen, Germany, without whose help this book would not have been possible.

TABLE of CONTENTS

Dedication		iii
Table of Contents		v
Preface to First Edition		ix
Preface to Second Edition		xi
Technical Note: TRM		xiii
Acknowledgements		xv
Part 0	BEFORE WE START	1
0	Sizing Traces	3
Part 1	BASIC CONCEPTS	5
1	Introduction and Historical Background	7
2	Materials used in PCBs	11
	2.1 Copper used in PCBs	12
	2.2 Dielectrics used in PCBs	15
3	Resistance and Resistivity	19
	3.1 Resistivity	19
	3.2 Resistance	21
	3.3 Thermal Coefficient of Resistivity (α)	22
	3.4 Measuring Resistivity	24
	3.5 Non-Destructive Measurements	26
4	Trace Heating and Cooling	29
	4.1 Trace Heating	29
	4.2 Trace Cooling	30
	4.3 Model of Trace Heating and Cooling	33

4.4 Role of Current Density 33
4.5 Measuring Trace Temperature 34
4.6 Temperature Curves 37

Part 2 EXPERIMENTAL STUDIES and SIMULATIONS 41
5 IPC Curves 43
5.1 Measuring the Data 43
5.2 IPC Curves 45
6 Thermal Simulations 53
6.1 Modeling Traces 53
6.2 The IPC Modeling Process 56
6.3 Sensitivities: Layout Parameters 62
6.4 Sensitivities: Material Parameters 70
6.5 Voltage Drop 77
6.6 Role of Current Density 77
6.7 Conclusions 77

Part 3 SPECIAL TOPICS 79
7 Via Temperatures 81
7.1 Background Information 81
7.2 Thermal Simulation 82
7.3 Experimental Verification 89
7.4 Experimental Results 91
7.5 Voltage Drops 93
8 Via Current Densities 97
8.1 Single Via 97
8.2 Multiple Vias 100
8.3 Multiple Vias and Turn 102
8.4 Conclusions 103
9 Fusing Currents, Background 105
9.1 W. H. Preece 105
9.2 I. M. Onderdonk 107
9.3 Derivation of Onderdonk Equation 110

10	Fusing Currents, Analyses	119
	10.1 "Fusing" Time and Temperature	120
	10.2 Assumptions and Cautions	120
	10.3 Simulation Models	120
	10.4 Simulation Results	121
	10.5 Short-time Effects	124
	10.6 Final Conclusions	127
	Supplemental Comment	128
11	Fusing Currents, Experimental Testing	129
	11.1 Problems in Predicting Fusing	129
	11.2 The Fusing Process	131
	11.3 Experimental Results	132
	11.4 Summary	136
12	Do Traces Heat Uniformly?	139
	12.1 Thermal Gradients on Traces	140
	12.2 Thermal Gradients on Narrow Traces	141
	12.3 Parallel Investigation	142
	12.4 Does Trace Thickness Matter	142
	12.5 What Causes Thermal Non-Uniformity?	143
	12.6 Conclusion	143
13	AC Currents	145
	13.1 Basic Models	145
	13.2 Experimental Verification	150
14	Thermal Effects, Right-Angle Corners	159
	14.1 Background	159
	14.2 Heating and Cooling Dynamics	160
	14.3 Experimental Verification	164
	14.4 Conclusions	166
15	Industrial CT Scanning	169
	15.1 Scanning Process	169
	15.2 Results	171

Part 4 APPENDICES 175

A1 Measuring Thermal Conductivity 177
A2 Measuring Resistivity 179
A3 Internal and Vacuum IPC curves fitted with equations 191
A4 Detailed set of equations for the curves 195
A5 Internal and Vacuum IPC Curves 197
A6 Current Density in Vias 201
A7 Fusing Current Reference 211
A8 Fusing Current Simulations 213
A9 Non-Uniform Heating Patterns 217

Abridged Index 219

About the Authors 221

PREFACE to First Edition

I wrote my first article on the relationship between PCB trace currents and temperature in 1998 [1]. In that article I compared two data sources: the then current IPC curves, and data from the 1968 issue of Design News. There was about a 40% difference in the results between those two sources that I was not able to explain.

Through the years I began to speculate that the difference might be related to thermal gradients on the traces. In the meantime, the IPC-2152 standard was published, suggesting an even bigger difference between the IPC data and the 1968 Design News data

Then, in late 2014, I opened up a dialog with Johannes Adam, in Leimen Germany. Johannes had authored a thermal simulation program called TRM, Thermal Risk Management, and I reasoned that that simulation package might finally help me answer the question about thermal gradients on traces. He graciously offered me a license for TRM, and we began a collaboration that lasted almost two years, and resulted in numerous papers and articles. [2]

In the first paper, we used TRM to validate the IPC 2152 curves (the IPC might suggest that we used the curves to validate TRM!) and the validation was wonderfully close. But the IPC curves are (necessarily) based on a set of strict procedures, so do not really represent typical board designs. In fact, it turns out they are almost a "worst case" scenario. We used TRM to run sensitivity analyses to see what the effects were of adding adjacent traces, underlying planes, or changing materials, etc.

We then looked the interesting question of what happens when you suddenly overload a trace --- how long does it take the trace to melt? And the results were pretty surprising when we compared them with the historical approach of dealing with such questions. But the real epiphany came when we simulated the temperature of a via. The results were completely contrary to the established wisdom of the time.

The via results were so surprising they almost literally cried out, "Show Me!" So I arranged for an experimental evaluation to confirm the simulation results. The experimental results were spot-on. But in the process we discovered there are uncertainties about additional issues with board design, namely the resistivity of copper on PCBs and the thermal properties of board dielectrics. This led to even further studies and analyses.

It was then that I came to the realization that we needed to put all this information together into a single publication --- and the idea for this book was born.

I have tried to make the information this book thorough and complete. Of course that is not possible. This is a very complex topic and there are a great many variables (and not a few uncertainties) involved. I have tried to open a door with this book, but there is still a lot to be discovered on the other side of that door.

I believe it is not possible to exactly thermally characterize a board design without actually testing the physical product. And even then, there is likely to be a lack of repeatability from one product to another. And I also believe it is becoming impractical to get close enough to the final results with charts, graphs, and equations. For a really optimal thermal analysis of a board design, a more complex simulation technique is required, one that requires some healthy computer analysis. And there is precedent for this in our industry. In the 1990's it became important (in some designs) to design impedance-controlled traces, and formulas were developed to support this need. But now we know that formulas are not accurate enough; field-solver techniques are usually required. The same thing is happening related to trace temperatures.

Notes

1. Brooks, Douglas, "Temperature Rise in PCB Traces," published on the UltraCAD website and printed in the "Proceedings of the PCB Design Conference, West," Miller Freeman, Inc., March 23-27, 1998.
2. See http://www.ultracad.com

PREFACE to Second Edition

Johannes and I continued our research after the first edition of this book was published. One of those research efforts was in the area of AC currents on traces. That research resulted in the addition of Chapter 13 **"AC Currents."**

Another research effort involved the addition of Norocel Codreanu to our team. Norocel is full professor at "Politehnica" University of Bucharest (UPB), Romania, Faculty of Electronics, Telecommunications and Information Technology, and currently the executive manager of the UPB university research center "Center for Technological Electronics and Interconnection Techniques" (UPB-CETTI). He was very instrumental in helping us with experimental verification of some right-angle corner heating Johanes and I were exploring. That effort resulted in a new Chapter 14, "**Thermal Effects, Right-Angle Corners"**

Then, as a result of several lectures we have given, Chapter 6 of the manuscript, **"Thermal Simulations,"** has been significantly expanded, including one entire new section. Another section was added to Chapter 7 on Voltage Drops Across Traces and Vias. Finally, we have included several, more modest, improvements in various other sections. Overall, we have added 36 pages to the publication, about 18% more material.

We are grateful for the reception the first edition received and hope this new edition makes the book even more valuable for our readers.

Technical Note: TRM

This book is written primarily for PCB designers, people who want to know "How big does my trace have to be in order to handle the current it is carrying." Naturally, this question is asked before the board is fully designed; it is too late to ask it *after* the board has been designed!

In this book I have made liberal use of the TRM (Thermal Risk Management) software tool to analyze thermal issues of and around individual traces, and the tool is very well suited to doing this. But this may lead the reader with the mistaken believe that that is the limit of what TRM can do. In fact, TRM is ***MUCH*** more powerful than this!

Dr. Johannes Adam originally conceived and designed TRM to analyze temperatures *across* an entire circuit board, taking into consideration the complete trace layout with optional Joule heating as well as various components and their own contributions to heat generation. Although the program could be adapted to the measurement of an individual trace, as we have done here, it was not originally conceived with that use in mind. So when I describe how to enter trace dimensions into the software using one of the menus (Figure 6.6), that omits the fact that an *entire board* can be entered into the software using, for example, the entire set of Gerber and drill files, a much simpler and more accurate (not to mention more complete) process.

I show many images of thermal diagrams the software produces relative to a single trace under analysis. But the figures on the next page illustrate how a similar thermal profile can be generated of an *entire* board, including components, under load. Imagine the power of this for a systems or package designer worrying about the thermal performance of an actual product in a difficult environment (such as the engine compartment of a car, or the confines of a very small enclosure!)

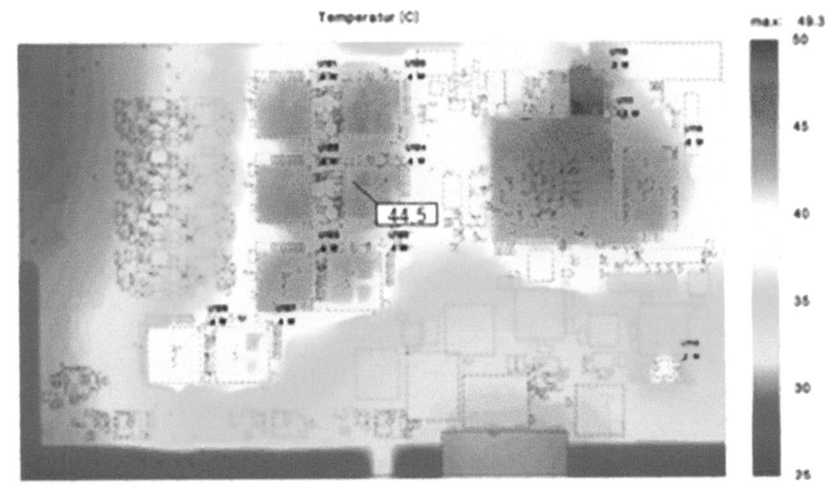

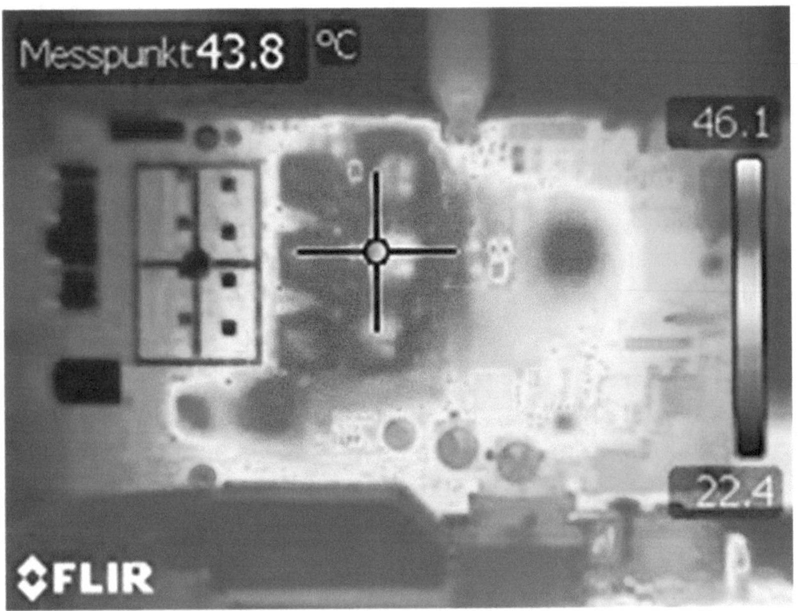

Simulation of actual board under load (top), and thermal image of the same board under operation (bottom).

You can learn more about TRM and all of its capabilities at:

http://www.adam-research.com

ACKNOWLEDGEMENTS

This book would not have been possible without the generous support of a lot of people. Lots of authors say this, but boy it is really true in this case.

In particular, I want to thank my longtime partner Dave Graves (now with Monsoon Solutions in Bellevue, WA) for helping prepare the final artwork for the various test boards. Adam Harris and Sarah Ackerman at C-Therm Technologies (Fredericton, New Brunswick) graciously measured the thermal conductivity of some board material to facilitate the via simulations. Fabio Visentin and Alejandro Golob of Jesse Garant Metrology Center, Windsor, Ontario, provided x-rays of a board and showed me how to analyze those x-rays. Nathaniel Peters of W. M. Keck Center for Advanced Studies in Neural Signaling provided valuable support in microscopy. Norocel Codreanu, of "Politehnica" University of Bucharest (Romania) provided support in looking at whether there are “hot spots” on traces (there are). And Scott Dau, a Seattle fireman and part time fire investigation instructor, provided valuable assistance in focusing in on localized hot spots on traces (Chapter 12) even at very low temperatures.

But a very special thanks to numerous people at Prototron Circuits, Inc. (of Redmond, WA and Tucson, AZ) who provided the test boards and also the microsectioning work and measurements for the various experimental tests. At the risk of accidentally leaving someone out, I want to particularly include Dave Ryder (president), Jerome Larez (Sales), Mark Thompson (Engineering Support) and Kevin Neumann (CAD, CAM, QA). Their support was fundamental in making this book possible.

But most of all, my collaborator on trace thermal issues, Johannes Adam (Leimen, Germany), was the person who started it all about 18 months ago. His interest, support, and encouragement (and tolerance) was what allowed me to start pursuing avenues that led to far more destinations than I ever expected. I could not have gone down this path without him.

Thank you all.

The success of this book (whatever it may be) belongs to all these people. And I hope there are no significant errors. But if there are any errors, they are the result of my imperfect understanding and in spite of everyone else’s efforts.

Acknowledgement, Second Edition Addendum:

Norocel Codreanu made a major contribution to this 2nd Edition. Norocel is full professor at "Politehnica" University of Bucharest (UPB), Romania, Faculty of Electronics, Telecommunications and Information Technology, and currently the executive manager of the UPB university research center "Center for Technological Electronics and Interconnection Techniques" (UPB-CETTI). He has done some earlier work in the area of trace heating, and provided invaluable support in the experimental verification of some of the models Johannes and I developed for the thermal investigation of trace right-angle corners. Chapter 14 could not have been written without his contribution.

Part 0

BEFORE WE START

0 SIZING TRACES

At its core, the purpose of this book is about how to size your trace to carry the required current. Here are the steps:

Step 1: You MUST know the maximum current (and perhaps the duty cycle) the trace will have to carry. It is the circuit design engineer's responsibility to determine this. If you don't know this, nothing else matters!

Step 2: Someone must determine the maximum allowable temperature increase of the trace. This is a policy variable; there is no right or wrong answer. Typically, it is the system design engineer who sets this policy. The policy will probably depend on the end market, the degree of reliability desired, and perhaps outside regulatory agencies.

Step 3: Given steps 1 and 2, select the trace size from the IPC 2152 tables. This will likely (but not necessarily) result in a worst-case size --- i.e. one larger than necessary.

Step 4: Determine the resistance of the trace. There are lots of calculators that can do this [1]. You can estimate it with the following formula:

$$R = .68 * L / (W * Th)$$

Where:
R = Resistance in Ohms at 20° C
L = trace length in inches
W = trace width in mils
Th = trace thickness in mils

Resistance will increase approximately 4.0% per 10° C rise in temperature.

Step 5: Calculate the voltage drop across the trace using Ohm's Law (V = current times resistance.)

Step 6: You now are in position to take one of three actions:

6a: Increase the trace size in order to lower the voltage drop across the trace (not a likely requirement.)

6b: Leave the trace size as determined.

6c: Decrease the trace size because there are reasons to know this trace has enough thermal margin to do so.

Choice **6c** is what this book is about. This choice does not depend on the circuit or the Power Distribution Network (PDN). It depends on the environment around the trace. And the impact of the environment is a very tricky thing to get our arms around. Unfortunately, there are no "cookie cutter" answers --- no formulas, no charts, no tables that can provide the answers. By the time you get through this book you will understand why that is.

Notes:

1. One such calculator is UltraCAD's free Wire Gauge Calculator, v3.0, available at http://www.ultracad.com.

Part 1

BASIC CONCEPTS

1 INTRODUCTION and HISTORICAL BACKGROUND

The first study of the relationship between trace currents and temperature is believed to have been done in 1956. It was published by the Natural Bureau of Standards as Report # 4283. There is a good summary of this effort contained in the IPC publication IPC2152 [1]. The experimental data were not very well controlled (because of limited resources), and the resulting charts, when published, were labeled "Tentative." The authors recommended that funding be provided for a more detailed, more carefully controlled study, but such funding was never forthcoming.

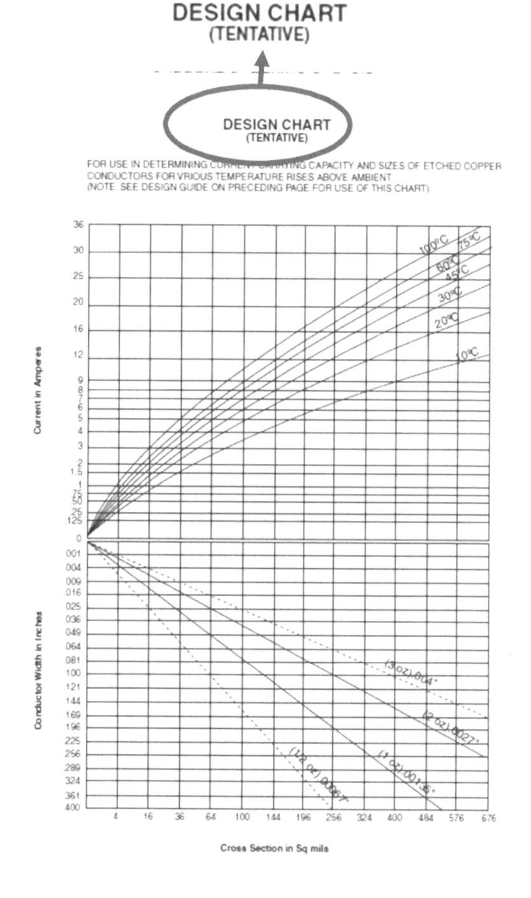

Figure 1.1
First known chart of trace current/temperature. Note the word "Tentative" in the title

Originally, there were two sets of charts, one for external traces and one for internal traces. The experimental data only applied to the external traces. The internal trace charts were derived by de-rating the external charts by a factor of two, on the expectation that the internal traces would not cool as well as the external traces would, and would therefore be hotter.

Through the years the charts were redrawn and republished, and somewhere along the line the word "Tentative" was dropped.

Apparently they were first published as part of MIL-STD-1495 in 1973. They were later published in a subsequent series of standards, culminating in MIL-STD-275E, published in 1984. Later still the charts were published as part of an IPC standard IPC-D-275 [2], and still later in IPC standard 2221 [3]. Copies of some of these source documents can be viewed on the Web [4].

These charts and standards were considered as the "Bible" for trace temperature measurements, in spite of their humble beginnings (and the fact that the assumption regarding internal traces would turn out to be quite incorrect!) They were used by most printed circuit board designers. In hindsight, the best thing they had going for them was the test of time. They apparently were appropriately conservative because few board failures were traced back to using them.

Between the initial publication of these charts in 1956 and 2009, I am only aware of one other study that tried to specifically characterize PCB traces with respect to current/temperature relationships [5].

Finally, the IPC helped sponsor a very thorough study on trace currents and temperature that was released as IPC-2152, in 2009 [1]. This is believed to be the best researched, best controlled, most thorough study ever made of trace currents and temperatures. The document is over 90 pages long and contains over 75 charts and tables. The results for external traces are (in my opinion) more evolutionary, than revolutionary. The IPC-2152 data result in currents approximately 25% lower than those shown in the original IPC-2221 set of curves, and they are much more detailed and complete. But the results for the internal traces are revolutionary. It turns out the internal traces cool better than do the external traces. That is because it turns out the board materials conduct heat away from the trace better than the air does. This is the one assumption the original researchers got very wrong.

I wrote my first article on the relationship between PCB trace currents and temperature in 1998 [6]. In that article I compared two data sources: the then current IPC curves and data from the 1968 issue of Design News. There was about a 40% difference in the results between those two sources that I was not able to explain.

Through the years I began to speculate that the difference might be related to thermal gradients on the traces. In the meantime, the IPC-2152 standard was published, suggesting an even bigger difference between the IPC data and the 1968 Design News data.

Then, in late 2014, I opened up a dialog with Johannes Adam, in Leimen Germany. Johannes had authored a thermal simulation program called TRM, Thermal Risk Management, and I reasoned that that simulation package might finally help me answer the question about thermal gradients on traces. He graciously offered me a license for TRM, and we began a collaboration that lasted almost two years, and resulted in numerous papers and articles. I discussed what happened next in the Preface.

In this book, for the first time, almost all the issues related to PCB trace temperatures are combined into a single publication. I start with basic theory of trace heating and cooling. This includes resistivity and how to measure it, and the thermal cooling capacity of board material, and how to measure that.

Then I look at how the IPC curves were developed, what they mean, and how they might be simulated with computer software. Then I look at sensitivity analyses, how many different variables are there and what happens when we vary them. This includes looking at the dynamics affecting via temperatures. And then I look at what happens when we catastrophically overload a trace. One thing happens if we overload it greatly (and it melts within seconds); something else happens if we slightly overload it and it takes from 30 minutes to over an hour to fail. And I provide experimental data to illustrate the various results presented in the chapters.

I have tried to make the information this book thorough and complete. Of course that is not possible. This is a very complex topic and there are a great many variables (and not a few uncertainties) involved. I have tried to open a door with this book, but there is still a lot to be discovered on the other side of that door.

I believe it is not possible to exactly characterize a board design without actually testing the physical product. And even then, there is likely to be a lack of repeatability from one product to another. And I also believe it is becoming impractical to get close enough to the final results with charts, graphs, and equations. For a really optimal analysis of a board design a more complex simulation technique is required, one that requires some healthy computer analysis. And, as indicated in the Preface, there is precedent for this in our industry. In the 1990's it became important (in some designs) to design impedance-controlled traces, and formulas were developed to support this need. But now we know that formulas are not accurate enough; field-solver techniques are usually required. The same thing is happening related to trace temperatures.

Notes:

1. For more information, see IPC-2152, “Standard for Determining Current Carrying Capacity in Printed Board Design,” August, 2009, Appendix A.7, p. 85. A copy of the original NBS chart is included there as Figure A-89, p. 86. Also see http://thermalnews.com/images/ThermalManagementLLC.pdf
2. ANSI/IPC-D-275, Design Standard for Rigid Printed Boards and Rigid Printed Board Assemblies, Figure 3-4, Page 10, IPC, September, 1991
3. IPC-2221, Generic Standard on Printed Board Design, 1998, superseded by IPC-2221A, Generic Standard on Printed Board Design, May, 2003, Figure 6-4, p. 41
4. See http://www.ultracad.com/thermal/thermal.htm.
5. “Printed Circuits and High Currents”, Friar, Michael E. and McClurg, Roger H., Design News, Vol. 23, December 6, 1968, pp. 102-107.
6. Brooks, Douglas, “Temperature Rise in PCB Traces,” published on the UltraCAD website and printed in the “Proceedings of the PCB Design Conference, West,” Miller Freeman, Inc., March 23-27, 1998.

2 MATERIALS USED IN PCBs

At first glance, a printed circuit board (PCB) is a pretty simple thing. There are one or more copper layers, etched into conducting patterns, and a substrate (or maybe substrate layers) to hold the copper. For most applications up until around the 1990's, that concept was sufficient. Board design was referred to as "connecting the dots." Then as circuit and component rise times got faster, individual traces began to act more like components, and problems like transmission line impedance control and EMI began to surface. This opened up a whole new world for some designers, especially those who had no background in basic electronics. This trend developed into a topic we now refer to as "signal integrity" related to board design.

As circuit components became faster and more complex, the power to operate them also began to increase. And this generated heat. Temperature and thermal management became a significant concern in some designs. More and more designers began to rely on standards like IPC-2152 to help determine how to appropriately size traces.

But stepping back for a moment, we might ask "Why does a trace heat up in the first place?" "Why does trace size matter?" After all, copper has a very low resistance to current, almost zero. And it conducts heat very well. Why isn't that sufficient?

In Chapters 2 through 4 we will look first at copper itself, and answer the question of whether copper is copper. That is, are there any relevant differences between types of copper used on a PCB? We will also look at the types of laminates PCBs are made from, and what parameters are important to us. Then, why do traces heat up? How do they cool? Where is the balance between heating and cooling? How do we measure trace temperature? And finally, is there anything that can change the balance between heating and cooling in the sense that a seemingly stable temperature becomes unstable and thermal "runaway" occurs?

2.1 Copper Used in PCBs

2.1.1 Copper-clad laminates: For surfaces where there will be conducting traces, the board fabricator typically purchases basic board dielectric material covered with a copper foil from one or more suppliers. The copper foil is typically either electrodeposited (ED) or it is rolled. In the ED foil manufacturing process, the copper is deposited on a conducting drum that slowly rotates in a copper sulfate solution. The drum is connected to a negative DC power supply and the positive supply side is connected to an anode in the solution. The slower the drum rotates, the thicker the plated copper foil (See Figure 2.1).

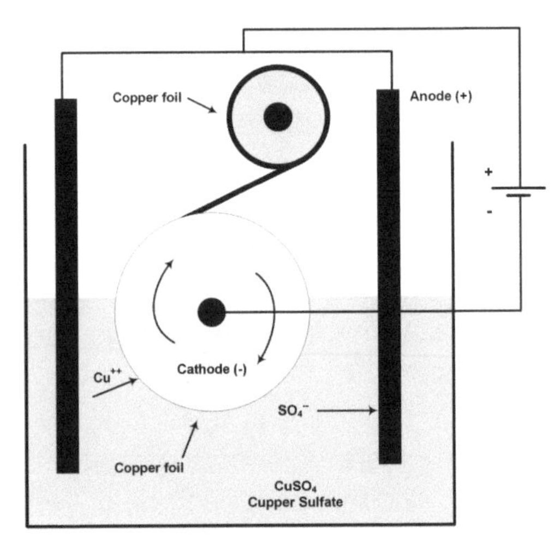

Figure 2.1
ED foil process

Rolled copper starts with a billet of copper alloy that is rolled between two rollers. This flattens the copper somewhat. It then goes through another pair of rollers, which flattens it further. Successive rolling operations result in the copper becoming thinner and thinner until it reaches the desired thickness (Figure 2.2).

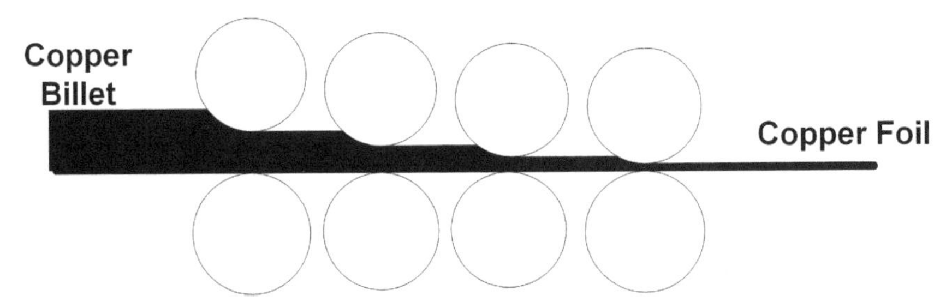

Figure 2.2
Rolled copper foil process

In both cases, the result is a sheet of thin copper foil. But there are two differences in these processes that may be of importance to the board designer. Rolled copper is flatter (depending on the roller surfaces) than ED copper. That is, the surface roughness is greater for ED copper. But the ED copper has two surfaces, and the "drum side" surface is smoother than the outside surface. So the outside (rough) surface provides a good surface for the copper to bond to the laminate without any intermediate processing. Rolled copper requires an intermediate step to "rough up" the surface in order to provide a good bond to the laminate. The practical importance of these differences are [1]:

1. Rolled copper is better for high-frequency applications (where surface roughness may have signal integrity implications.)
2. ED copper is better for bonded assemblies.
3. ED copper has a lower resistivity (after bonding to a laminate) (1.62-1.66 uOhm-cm) than rolled copper (1.74-1.78 uOhm-cm.)

The copper foil process is pretty well-controlled and tolerances are good. Foil thickness is typically held within 10% or better. Resistivity is reasonably constant across the surface of the foil. While there is opportunity for contamination to occur between the foil and the board (which may impact some thermal properties), this should be limited to a few spots on the board and should not significantly impact the thermal performance of the board.

2.1.2 Copper plating manufacturing step: The board fabricator starts with the basic board material and builds up a finished board. Inner layers will go through etching processes to etch the trace layer pattern into the copper, and for any blind and/or buried vias, may go through a separate plating step. Then the layers are assembled and bonded together. There is a point where the layers of the board have been fully built up, and the board (really manufacturing panel) exists as an assembly with solid copper foil on the top and the bottom and holes drilled for the vias.

From here, the process can take one of two directions:

1. The top and bottom patterns can be etched into the copper, and then board (and traces) plated with copper to cover the walls of the vias (this process is referred to as pattern plating), or
2. The entire board area (top and bottom) can be plated with copper and then the trace patterns etched into the combined foil plus plating layer (referred to as panel plating.)

In both cases the outside layer copper foil is covered with plated copper. This plating step is very similar to the plating step for copper foil --- the board is dipped into a copper solution and the copper is electroplated onto the board. Then the board is withdrawn from the solution (Figure 2.3). In the first example the trace patterns are etched before plating, in the second example the pattern is etched after plating. In both cases the traces consist of copper foil covered with plated copper. There may be minor differences in the thickness and shape of traces etched in these two manners, and these differences might be important to signal integrity engineers --- they may have an impact in trace impedance calculations for example. But they are unlikely to significantly impact the trace's thermal characteristics.

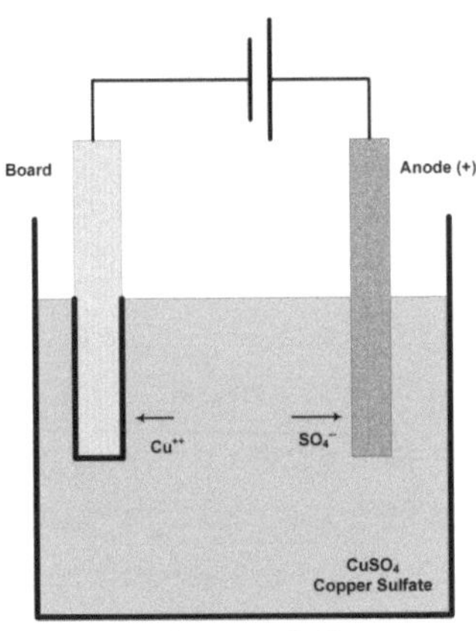

Figure 2.3
Plating of board

But there are other things here that might be important to us. First, and perhaps most important, is that this plating process itself is difficult to control. Different areas of the board may plate to significantly different thicknesses. Thickness variation of up to 50% should not be unexpected and there is anecdotal evidence on the Web that plating thickness might vary by as much as 100% around the board. This variation can be controlled somewhat by the fabricator, but that might involve additional steps, involve modification to the board design, and might involve additional expense. Such efforts might reduce the variations in thickness, but variations as much as 20% may still occur. As we will see below, trace thickness has a significant impact on trace resistance, which therefore has an impact on trace heating.

Second, plating may impact resistivity (see the next section.)

Third, the many processing steps open up the possibility of contamination, especially between the foil and plated copper layers. Such contamination, if it exists, may impact the thermal characteristics of the board, especially at elevated temperatures.

Finally, the pattern etching process has a tolerance that applies to the finished width of the trace. If the trace width varies along the trace, this may impact the thermal characteristic of the board, and if it does, where "hot spots" may exist along a trace. The narrower a trace, the more important this tolerance is.

2.1.3 Copper Resistivity: The resistivity of copper is published in a very large number of places [2]. Resistivity is typically specified as 1.68 uOhm-cm for "pure" copper and 1.72 uOhm-cm or more for a copper alloy (annealed copper.) The problem is complicated by the number of copper alloys there are.

There are hundreds of copper alloys defined. There is a classification system called the "Unified Numbering System for Metals and Alloys" (UNS for short). Copper alloys have UNS numbers starting with Cxxxxx [3]. "Coppers" (UNS C10100 to C15999) have a purity => 99.3%, and High-copper alloys (UNS C16000 to C19999) have a purity > 96%. Selecting random UNS alloys in reference [3] in the range of C10100 to C14500 results in a range of resistivities from 1.69 to 1.86 uOhm-cm.

If there are any impurities in the copper, the resistivity will change. An interesting chart in a paper titled "High Conductivity Coppers" [4] shows the approximate effect of impurities on mass electrical resistivity [5]. Changes in resistivity depend very much on the impurity and the concentration, but the variation can be 25% with even small changes of some impurities. (See Chapter 3 for more on copper resistivities.)

Summary: For trace current/temperature considerations, the following copper trace parameters are important:

1. Variations in finished trace thickness
2. Variations in resistivity
3. Points where contamination may exist between the foil and the board material, or between the plated copper and the copper foil parts of a trace.

2.2 Dielectrics used in PCBs

A search of "PCB dielectrics" on the Web results in a significant variety of laminates a board designer might choose from. The most common dielectric is standard FR4, but even that category has many variants. Recent new material choices have arisen because of signal integrity needs. High-frequency materials have been developed for their dielectric coefficient stability with frequency and/or to minimize their dielectric losses. But none of that matters for our purposes. There are four parameters that matter from a current/temperature standpoint, and only one of those matters at normal temperatures.

2.2.1 Thermal conductivity (Tcon or k): Dielectrics conduct heat in two "directions." One is in the x-y plane, parallel to the surface layer, and therefore parallel to the traces. This is sometimes called the "in-plane" conductivity. The other is in the z-axis, or perpendicular to the surface. This is sometimes called "through-plane." The measure that reflects this thermal conductivity is called the "thermal conductivity coefficient," sometimes referred to as Tcon.-x and Tcon-z, respectively. Typical values for Tcon-x might be 0.5 or .6 and for Tcon-z might be 0.3 to 0.5 (compare with copper at 386). Material specifications frequently fail to include Tcon values, and if they do they are unlikely to specify whether the value applies to the in-plane or through-plane direction. Units of Tcon are Watts/meter-degree Kelvin, or W/mK.

There are now tools available to measure thermal conductivity at a relatively modest cost. One such tool is offered by C-Therm Technologies in New Brunswick, Canada (See Appendix 1).

This parameter is important to us because it directly relates to the dielectric's ability to conduct heat away from the trace.

Three other material parameters are important to us if the trace reaches a moderately high temperature (say higher than 100° C.)

2.2.2 GlassTransition Temperature (Tg): (See IPC Test Method 2.4.24) [6]. Glass transition temperature is where the polymer board material transitions from a hard, glass-like material to a soft, rubbery material. (This is *not* to be confused with melting.) It is really a range of temperature over which this transition occurs, and Tg is defined as the center of that temperature range. During the transition, the board material Coefficient of Thermal Expansion (CTE) increases, which means the board itself expands. Since the board material is somewhat constrained by copper traces and planes in the x-y direction, the dielectric expansion tends to occur in the z-axis, sometimes putting mechanical stress on nearby vias. In extreme cases, via conductivity may be lost as a result of this expansion.

Since the dielectric material is changing form, and since the dimensions are changing, therefore the thermal conductivity is changing. This impacts the cooling characteristic of the material. Tg of different materials typically ranges from around 90° C to about 190° C . A board is perfectly usable at temperatures above Tg, and the effects of Tg are reversible.

2.2.3 Decomposition Temperature (Td): Decomposition temperature is the temperature at which a PCB material chemically decomposes (i.e. the material loses at least 5% of mass). This results in changes in material properties and therefore in a change in how the material can conduct heat away from the trace. Td is important because the changes that result are irreversible. (This is a consideration for assembly processes, especially flow soldering operations.)

Fortunately, typical Td specifications are relatively high, often in the range of 350° C.

2.2.4 Time to Delamination (T260/T288): There is a test known as "Time to Delamination" and is defined in IPC Test Method 2.4.24.1 [7]. It is sometimes referred to on specification sheets as "Thermal Resistance." A test sample is incrementally raised in temperature 10°C/min to 260°C (or 288°C) and then is held at that temperature. When a destructive event occurs (such as delamination, cracking, moisture release, stress relaxation, decomposition or a sudden movement) a sensor detects the change. The time in minutes to the event at 260°C (or 288°C) is the time to delamination. The delamination event is destructive and irreversible. Times for different laminates can vary between a few minutes to an hour or more.

This is relevant for us in the following situation. Suppose we have a trace carrying a very high current that needs supplemental cooling (forced air flow or heat sink.) Then suppose something happens that for some reason degrades the supplemental cooling and the trace temperature rises to the delamination temperature. The temperature may hold steady for a while, but if the board begins to delaminate, all bets are off. Once delamination starts, the conductive cooling through the board material is disrupted and the temperature will start increasing, often resulting in a thermal "runaway" condition and subsequent melting of the conductor (see Chapter 11).

Summary: For trace current/temperature considerations, the following laminate material properties are important:

1. Thermal conductivity (Tcon or k): relates to how well the material carries heat away from the trace.
2. Glass Transition Temperature (Tg): temperature at which the polymers change from hard, glassy like appearance to softer plastic-like appearance (reversible).
3. Decomposition Temperature (Td): Temperature at which the board material begins to decompose (irreversible).
4. Time to Delamination (T260/280): Time for the board to begin delaminating (irreversible and very destructive). This event can lead to thermal runaway.

Notes:

1. "Copper Foils for High Frequency Materials," Rogers Corporation, 2015, 100 S. Roosevelt Avenue, Chandler, AZ 85226
2. See, for example, http://hyperphysics.phy-astr.gsu.edu/hbase/tables/rstiv.html
3. See, for example, http://www.matweb.com/search/SearchUNS.aspx
4. "High Conductivity Coppers for Electrical Engineering," May 1998, Copper Development Association, www.cda.org.uk , Figure 6, page 47. Paper can be downloaded here: http://copperalliance.org.uk/docs/librariesprovider5/pub-122-hicon-coppers-for-electrical-engineering-pdf.pdf?sfvrsn=2
5. Mass resistivity = volume resistivity * density
6. See IPC Test Manual IPC-TM-260 Test Method 2.4.24, "Glass Transition Temperature and Z-Axis Thermal Expansion by TMA. . A complete set of test manuals and test methods can be freely downloaded from http://www.ipc.org/test-methods.aspx
7. See IPC Test Manual IPC-TM-260 Test Method 2.4.24.1, "Time to Delamination (TMA Method)". A complete set of test manuals and test methods can be freely downloaded from http://www.ipc.org/test-methods.aspx

3 RESISTIVITY and RESISTANCE

3.1 Resistivity

Resistivity is a measure of resistance of an element to the flow of current. So let us first understand the nature of current. To do that, we must look at atomic structure. Figure 3.1 is a simple model of an atom. An atom has a nucleus consisting of protons and neutrons, surrounded by electrons. (We know that atoms are more complicated than this, but this is all we need for now.) The number of protons defines the element. Copper has 29 protons. An atom has the same number of electrons (negatively charged) as it has protons (positively charged). So in all but very special cases, atoms are electrically neutral.

Electrons "fly" around the nucleus in energy shells. We don't know how to draw energy shells, so we draw them for convenience as concentric circles. Thus, the model looks like a sun with planets rotating around it, giving it the name "planetary model." Now these shells have capacities. The inner shell has a capacity of two electrons. The next shell has a capacity of 8 electrons. It tends to get more complicated from there, but it is sufficient to understand that much.

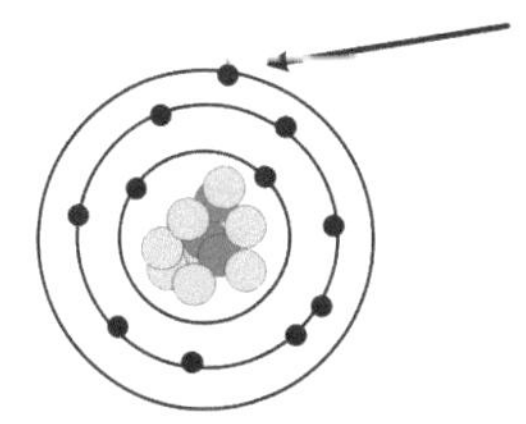

Figure 3.1
Planetary model of an atom.

Now here is the punch line: Atoms with a single electron in their outermost shell tend to have the lowest resistance to current flow. The reason is that that electron is not very tightly coupled to the nucleus. If there are more electrons in the highest energy shell, they tend to be more tightly coupled to the nucleus. Loosely held electrons are the easiest to displace. Copper (as well as silver and gold) has a single electron in it's outer shell. So if we apply a force (think voltage) across a copper conductor, that force can dislodge the electrons in the outer-most shell from one atom to the next.

Figure 3.2 is a model of a copper trace. Imagine a voltage (force) applied across the conductor by a battery. The battery "pushes" an electron onto the trace. It is attracted by the opposite charge at the far end of the trace. An electron must therefore leave the trace at the opposite end because the trace (and the atoms making it up) must remain electrically neutral [1]. That movement (shift) in electrons is the definition of current. The injected electron travels along the trace, but in doing so it collides with an electron that is already there. The injected electron may stay in place, replacing the electron it collided with, and the displaced electron continues forward under the force of the voltage. This forward motion of the electrons, and their collisions, continues at a very high rate. (One Amp of current is defined as the movement of 6.25×10^{18} electrons passing across a surface in one second. That's a lot of electrons!)[2]

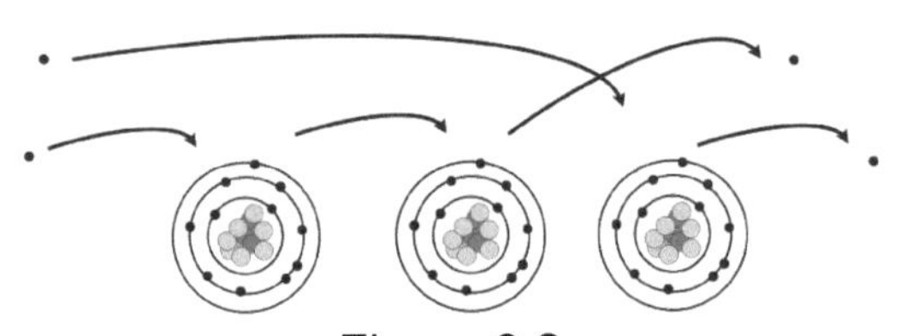

Figure 3.2
Model of a conductor.

Since the single electrons in the outer shell are loosely held, it does not take much energy to displace them. But the required energy is not zero. There is a small amount of energy that is dissipated each time one electron collides with another and the second electron is displaced. That energy loss is the nature of resistance and resistivity.

Now we intuitively know that silver, copper, and gold are excellent conductors of electricity. They have three characteristics of interest here:

1. They have the lowest values of resistivity of all elements
2. They all have a single electron in their outer shell
3. They all are solids at room temperature.

Their reported resistivities are [3]:

Silver	$1.6x10^{-8}$ Ohm-m = .63 μOhm-in
Copper	$1.7x10^{-8}$ Ohm-m = .67 μOhm-in
Gold	$2.2x10^{-8}$ Ohm-m = .87 μOhm-in

Note: Units of resistivity are Ohm-length

For comparison, Silicon has a resistivity from 0.1 to 60 Ohm-m and has four electrons in its outer shell. Glass has a resistivity of 1.0×10^9 to 1.0×10^{13} Ohm-m. Glass is an excellent insulator.

The resistivity reported above is the resistivity of "pure" copper. If there are any impurities present (as with a copper alloy, for example), electrons will also collide with them, dissipating more energy and the resistivity will increase. So any contamination of the copper used in a trace will increase the resistivity of the copper and the resistance of the trace.

3.2 Resistance

Resistivity is a parameter associated with an element (such as copper) or a combination of elements (such as a copper alloy.) Resistance is a property of an element that has a specific shape (such as a conductor.) The resistance of a conductor is found by dividing the resistivity by the cross-sectional area of the conductor and then multiplying by the length, or:

[Eq. 3.1] $$R = (\rho/A)*L$$

Where R = resistance in Ohms
ρ = resistivity in Ohms-length
A = cross-sectional area in square units
L = length in units

So if we know the resistivity of copper (say 0.67 uOhm-in), then we would calculate the resistance of a 0.5 Oz. 100 mil wide, 6" long trace as:

$$R = ((0.67*10^{-6})/(.00065*0.1))*6 = 0.0618 \text{ Ohms}$$

Conversely, if we know the resistance of a trace (because we measured it directly or calculated it from Ohm's Law), then we can calculate the resistivity of the trace material by:

[Eq. 3.2] $$\rho = R * A/L$$

3.3 Thermal Coefficient of Resistivity (α)

Resistivity also increases with temperature. The reason is that temperature is motion. That is, as the temperature increases, atomic and molecular motion increases. (Absolute zero, -273 °C, is when all atomic motion ceases.) When the atomic motion increases, there are more collisions among the electrons carrying the current, resulting in an additional loss of energy. Therefore, resistivity values are always specified at a particular temperature, typically 20 degrees C.

The change of resistivity of copper (indeed almost any element) is predictable. The parameter, *thermal coefficient of resistivity* (α) is the relative change in resistivity associated with a given change in temperature (Equation 3.3).

$$\alpha = \frac{\Delta\rho / \rho}{\Delta T}$$

[Eq. 3.3]

Where: α = thermal coefficient of resistivity
ρ = Resistivity
Δρ = Change in resistivity
ΔT = change in temperature

Since resistance is simply resistivity modified by the geometry of the conductor, it follows that we can substitute resistance for ρ in Equation 3.3 and get Equation 3.4:

$$\alpha = \frac{\Delta R / R}{\Delta T}$$

[Eq. 3.4]

α itself is a function of temperature. It is correctly stated as a value at some specific temperature (most commonly 20° C.) So, Equation 3.4 leads to the fundamental relationship between resistance and temperature:

[Eq. 3.5] $R(T) = R(T_o)*(1+\alpha_o*(T - T_o))$

Where: T = temperature of interest
T_o = reference temperature
$R(T)$ = Resistance at the temperature of interest
$R(T_o)$ = Resistance at the reference temperature
α_o = Thermal coefficient of resistivity at the reference temperature.

Equation 3.5 applies equally well to resistivity, ρ:

[Eq. 3.6] $\rho(T) = \rho(T_o)*(1+\alpha_o*(T - T_o))$

Where: T = temperature of interest
T_o = reference temperature
$\rho(T)$ = Resistivity at the temperature of interest
$\rho(T_o)$ = Resistivity at the reference temperature
α_o = Thermal coefficient of resistivity at the reference temperature.

Matula [4] has a table of copper resistivities as a function of temperature. The figures are for at least 99.99% "pure" copper. This goes up to and including the melting point of copper and then for liquid copper. From his data we can derive the implied thermal coefficient of resistivity (from ^{2}C using Equation 3.3. Table 3.1 is derived from Matula's data. Figure 3.3 is a graph of resistivity vs temperature.

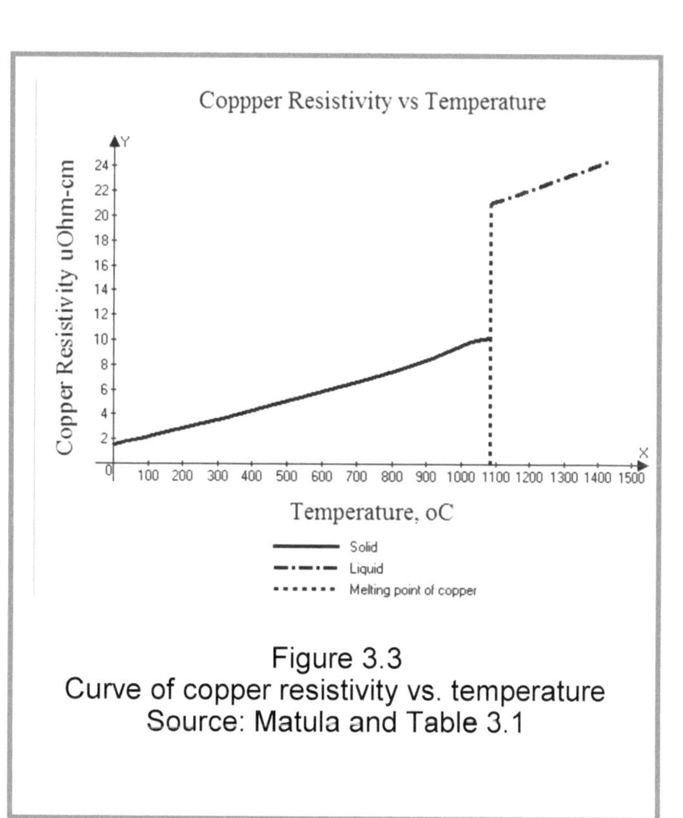

Figure 3.3
Curve of copper resistivity vs. temperature
Source: Matula and Table 3.1

3.4 Measuring Resistivity

Suppose we wish to measure the resistivity of a copper trace. We can do so relying on Equation 3-2. But there are some significant barriers to doing so.

First, we need to determine R, the resistance of the trace. We can do so a couple of different ways. One is to measure it directly with an Ohmmeter. There are precision meters available that can do this with a high enough precision. Or, we can apply a known current through the trace and measure the voltage (or vice-versa) and calculate resistance using Ohm's Law, R = v/i. This can also be accomplished with a reasonably high precision, especially under controlled conditions.

Second, we need to be able to measure or calculate the cross-sectional area of the trace. This means knowing (or measuring) the trace width and thickness. It is important to recognize that *this cannot be done by using the nominal dimensions of the design*. There is far too much tolerance in the dimensions to be able to do this. If there is copper plating applied over copper film, trace thickness can vary from nominal by as much as 50% (or even more). The photo etching process can have a tolerance of at least one or two mils.

Third, the trace length would seem to be measureable, but there are possible problems here, too. If there are pads at the ends of the trace, their area may need to be considered, especially if the pad is approximately the same size as the trace. Only in the case where the pad is signifi-

Temp, °C	Resistivity uOhm-cm	Implied α From 20 °C
0	1.541	
20	1.676	
27	1.723	.0040
77	2.061	.0040
127	2.400	.0040
227	3.088	.0041
327	3.790	.0041
427	4.512	.0042
527	5.260	.0042
627	6.039	.0043
727	6.856	.0044
827	7.715	.0045
927	8.624	.0046
1027	9.950	.0047
1084	10.169	.0048
1084*	21.01	
1127*	21.43	
1227*	22.42	
1327*	23.42	
1427*	24.41	

Table 3.1

Resistivity as a function of temperature.

(* = liquid copper)

(Source: Derived from Matula's data [4].)

cantly larger than the trace (in which case its resistance is significantly lower than that of the trace) can we perhaps be able to ignore the pad. If the pad's dimensions are different than that of the trace, the calculations may become significantly more difficult.

Finally, the temperature may be a factor. If the trace is at an elevated temperature and we wish to end up with the resistivity at 20 °C, then we need to do the type of conversion related to Equation 3.5. This implies we know *both* the elevated temperature and α, neither of which can be casually assumed.

Often, when we want to know trace dimensions precisely, we microsection a trace to take the measurements. Figure 3.4 shows two typical microsections from a test board. The one on the left is for measuring a via wall thickness, the one on the right for measuring the width and thickness of a trace. The problem with microsectioning is that (a) it is destructive, and (b) the typical microsection measurement tolerance is around 0.1 mil. That is about a 10% or 15% measurement error for a half-ounce trace! Furthermore, microsectioning will give you the dimensions of the trace *at that point*. But the dimensions might change as we move along the trace. Microsectioning cannot help us determine the extent of such variation.

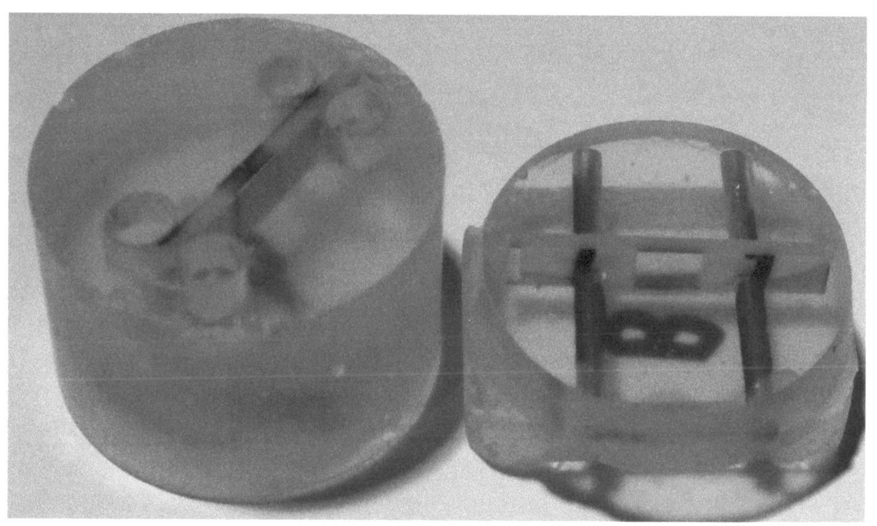

Figure 3.4
Typical microsection samples from a test board.

It is this author's opinion that with today's technologies, it is impractical to try to measure the resistivity of a typical PCB trace without the use of advanced and sophisticated measurement tools in a well-controlled lab. We simply cannot do it precisely enough. There are numerous anecdotal examples reported on the Web of people trying to measure resistivity numerous ways, with results ranging from 1.6 to 2.7 µOhm-cm, but it is believed all of them suffer from the uncertainties outlined above.

3.4.1 Resistivity investigation: With Prototron Circuit's help [5], a special test board was prepared to measure the resistivity of copper foil, plated copper, and their combination. The special design of the test board enabled the calculation of trace resistivity with reasonable precision. What was determined was that the resistivity of the copper plating was almost exactly that of pure copper --- just what one might expect.

The details of that investigation are reported in Appendix 2.

3.5: Non-destructive Measurements

There is an emerging technology, Computed Tomography (Industrial (CT) Scanning) that might make non-destructive trace measurements more practical. The resolution is not quite high enough today to be useful in these kinds of analyses, but I am optimistic that it might be in the not too distant future.

I had the opportunity to send a section of a board to of Jesse Garant Metrology Center, Windsor, Ontario, for x-raying. The results of that investigation are reported in Chapter 15.

Notes:

1. From this much we can infer two of the most important laws in electronics: (a) current must flow in a loop, and (b) current is constant everywhere in the loop. See Brooks, Douglas G., PCB Currents, How They Flow, How They React, Prentice Hall, 2013, Chapters 1 and 4.
2. Some students get confused by this model and how it squares with the fact that signals propagate at the speed of light. There are *two* velocities here. One is the speed of a single, particular electron. It travels at a speed called the "drift" velocity – most of us can walk faster than that. The other is the speed with which the transfer of energy takes place among the atoms that make up the trace. That transfer of energy happens at the speed of light. Again, see note 1, above.

3. Resistivity values for all the elements are readily available on the Web and in various handbooks.

4. R. A. Matula, Purdue University, "Electrical resistivity of Copper, Gold, Palladium, and Silver," as reprinted In the Journal of Phys, Chem Ref Data, Vol 8, No. 4, 1979. See especially Table 2, p. 1161.

5. Prototron Circuits, Redmond, WA. And Tucson, AZ., graciously provided some test boards for this investigation.

4 TRACE HEATING and COOLING

Traces heat because of I^2R power dissipation, increasing the temperature. This heating is above the initial condition of the trace [1]. The trace's resistance is inversely proportional to the trace's cross-sectional area. Traces cool because of conduction, convection and radiation, lowering the temperature. A trace's cooling characteristics are related to the surface area of the trace at any point along the trace. If the power is changing with time, then the total energy being dissipated in the trace will be changing with time. The cooling may also change with time. A stable temperature is reached when heating equals cooling over a stable time frame (see Figure 4.1).

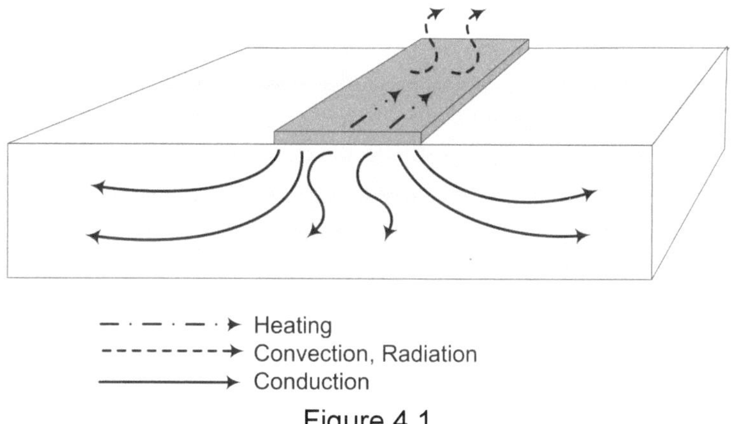

Figure 4.1
Trace heating and cooling dynamics.

4.1 Trace Heating

We noted in Chapter 3 that traces have resistance --- not much, but resistance is not zero. Therefore, a current through a trace will produce a voltage across the trace (and vice versa). Power is voltage times current:

[Eq. 4.1] Power = V * I (Units are watts)

By Ohms Law, V = I*R, so we can, by substitution, rewrite Equation 4.1 as:

[Eq. 4.2] Power = I^2R (Units are watts)

We often take the term "power" very casually, so let's take a moment to be clear about the difference between *power* and *energy*.

4.1.1 Power and energy: *Energy* is the *amount* of work done. Power is the *rate* at which we do it. A 100 watt light bulb dissipates power at the *rate* of 100 watts (per unit time). The total amount of *energy* it dissipates in an hour is 100 watt-hours. We pay for *energy* usage in our homes (not power usage), typically based on a charge per kWh – kilowatt-hour.

4.1.2 Trace Heating: The I^2R power dissipation in a trace is a "point" concept. It reflects the amount of energy being dissipated at that point in time, and at that point along the trace. It is at that point in time because if voltage (for example) is changing, the power will also be changing. It is at this point along the trace because the trace parameters may change at any point along the trace. The total energy dissipated in the trace is the integral of the power dissipated at a point along the trace times the time it is being dissipated, integrated over the length of the trace.

Resistance, as we covered in Chapter 3, Equation 3.1, is resistivity times length and divided by cross-sectional area. If the area changes *at any point* along the trace, the power dissipation *at that point* will also change. Similarly, if resistivity changes *at any point* along the trace (perhaps because of contamination of the copper), then the power dissipation *at that point* will also change. Thus, a trace does not necessarily heat uniformly along the trace. There may (and probably will) be some places hotter than other places along the trace, particularly at higher temperatures. (See Chapter 12.)

4.2 Trace Cooling

Traces on the top and bottom layers of a board have roughly half their surface area exposed to the air, and half in contact with the board material [2]. These traces cool by conduction, convection, and radiation. Internal traces are surrounded by board material and cool almost exclusively by conduction.

The ability of the board material to conduct heat away from the trace is largely determined by the thermal conductivity of the material. This conductivity can proceed in two directions: (a) away from the trace in the x,y plane (parallel to the trace and to the surface of the board) and in the z-axis (perpendicular to the trace and to the surface.) These directions are often referred to as "in-plane", and "through plane" respectively. Different board materials will have different thermal conductivity coefficients. These coefficients may or may not be included in the material specifications from the material supplier. If the coefficients are not available they may be estimated, but thermal modeling becomes more difficult.

Thermal conductivity coefficients are typically in the range of 0.3 to 0.8 W/mK (watts per meter degree K). Higher thermal conductivity means the board conducts heat better.

The transfer of heat away from the board into the air is determined by the "heat transfer coefficient (HTC)" [3]. There are two components of HTC, convection and radiation. The convection component of HTC can be a little tricky to estimate. First, it is a function of the temperature difference between the trace and the air (e.g. it would increase if the air were cooled, and might decrease as the air around the trace heats up.) Thus HTC is not necessarily constant during an investigation as the trace temperature increases. Second, it is a function of the velocity of the air (i.e. if a breeze is blowing.) Thus, HTC increases if the trace is cooled by a fan. In any experimental investigation it is important to take measurements in still air and to wait long enough for the temperature differences between the trace and the surrounding air to stabilize.

The heat transfer characteristics of convection (through the air) and radiation (into space) are approximately equal in laboratory conditions. Thus, we would expect surface traces operated in a vacuum would cool about 50% less effectively (by convection and radiation) than they would in a non-vacuum condition. That is, the HTC coefficient would be about 50% of what it would be in a non-vacuum situation. That means we would expect traces in a vacuum would be hotter than traces not in a vacuum. And in fact we see that in the IPC data. We will see later that that is confirmed in simulation models.

It turns out that the thermal conductivity of board materials is more efficient than the heat transfer through convection and radiation into the air. Therefore, internal traces (surrounded by board material) cool *better* than traces on the surface. This was pretty much unrecognized until IPC-2152 was published in 2009. Before that, the standard convention was to derate internal traces by 50% on the assumption that they cooled much less effectively than traces cooled by air.

My hypothesis is that the fiberglass in dielectrics conducts heat better than does the resin. Therefore, a board material with layers of fiberglass embedded in resin tends to cool better in the horizontal (or in-plane, or x,y plane) direction than it does in the vertical (or through plane, or z-axis) direction. And the cooling flow tends to follow the fiberglass layers (Figure 4.2). As a result, board materials are "anisotropic," that is they conduct heat better in one direction than in another. (See Appendix 1 for more on thermal conductivity and how it is measured.)

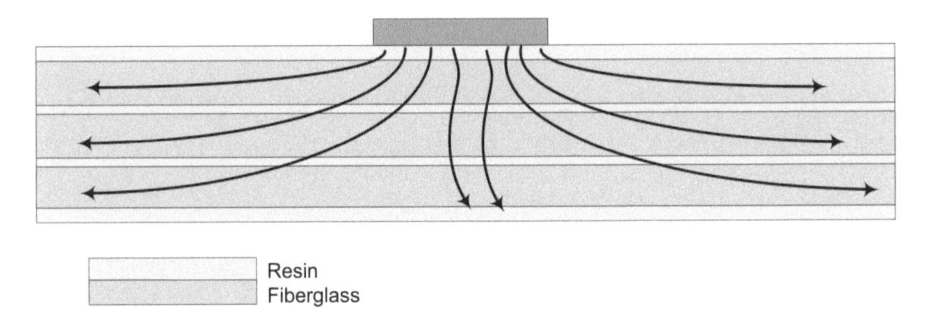

Figure 4.2
Cooling tends to follow the fiberglass layers in dielectrics.

Finally, narrower traces cool (relatively speaking) better than wider traces. The reason is illustrated in Figure 4.3. During the cooling process, heat must conduct, or flow, out from under a trace. As shown in the figure, the transfer path is significantly longer for a wider trace than it is for a narrower one. Furthermore, since the cooling path is shorter at the edge of the trace than it is at the midpoint of the trace, there is typically a thermal gradient from the hotter centerline out to the cooler edges. See more on this in Chapter 12.

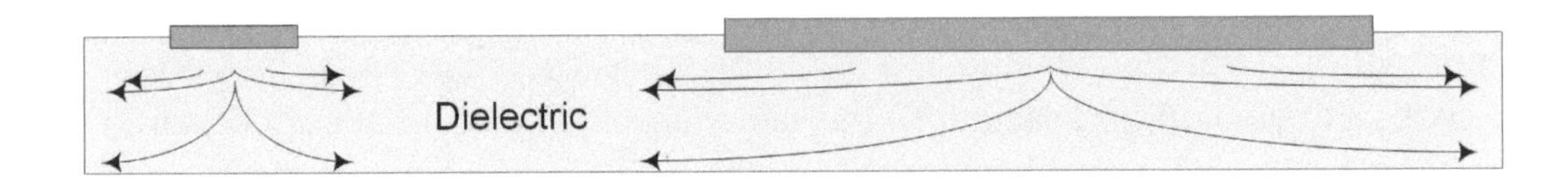

Figure 4.3
Narrow traces have a shorter cooling path
compared to wider traces.

This is evident, also, in simulation results. Figure 4.4 illustrates the thermal "plume" around a trace. Both traces in the figure are carrying enough current to heat them to about 70°C. The figure illustrates how the temperature gradient spreads out on the surface of the board away from the trace. In this model, the plume around the 200 mil (5 mm) trace extends out roughly 10 mm each side of the trace, or about 4x the trace width. The plume around the 20 mil (0.5 mm) trace extends out roughly 7.5 mm each side of the trace, or about 30 times the trace width. This illustration applies to a model where the temperatures have stabilized. We will see later that in situations where the trace temperature rises very quickly (as in a fusing situation) the difference is even more dramatic.

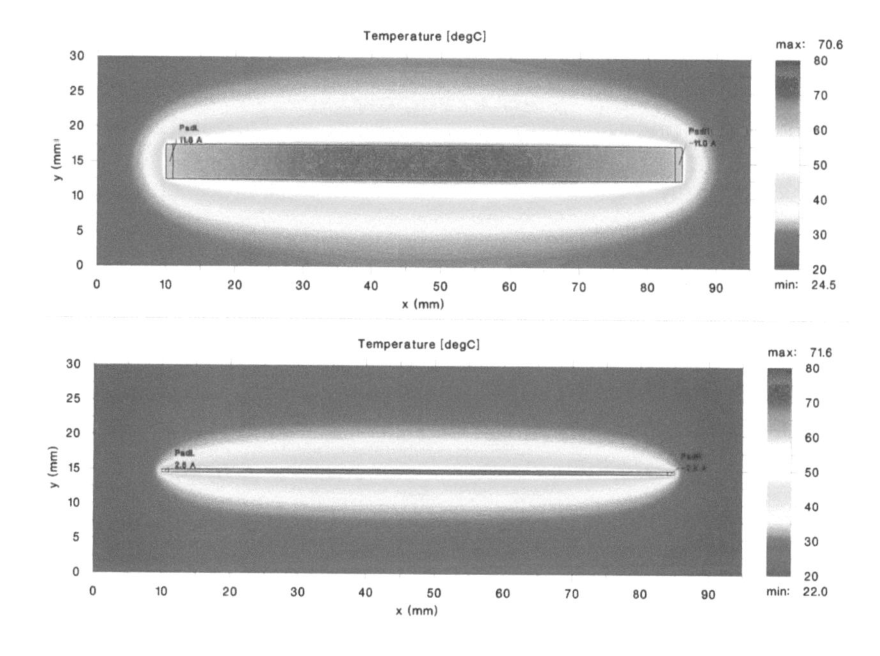

Figure 4.4
Thermal plumes around a 200 mil wide trace (top)
and 20 mil wide trace (bottom)

4.3 Model of Trace Heating and Cooling

A constant trace temperature occurs when heating equals cooling. Heating is a function of power dissipation in the trace, I^2R. The cooling relationship is complicated, but is related, in part, to trace surface area, or to width plus thickness, (W + Th). So, the change in temperature will be proportional to I^2R and inversely proportional to (W+Th), or

$$\Delta T \propto \frac{I^2 R}{(w + Th)}$$

[Eq. 4.3]

Where:

- ΔT = Change in temperature
- I = current
- R = trace resistance
- Th = trace thickness
- w = trace width

4.4 Role of Current Density:

There are some who believe that trace temperatures can be correlated to current density (A/in^2). Actually there is no direct relationship between current density and temperature. Here are two examples to illustrate the point.

Case 1: Assume two traces, A and B, are each carrying the same current. They each have the same cross-sectional area. Therefore the current densities are identical. But trace A is wide and thin and trace B is narrow and thicker. Trace A will be cooler because it has more surface area from which to dissipate heat.

Case 2: Assume two traces, A and B, are each carrying the same current and have identical cross-sectional areas and form factors. The current density will be identical in each trace. The resistivity of the conductor material making up trace A is higher than the resistivity of the conductor material making up trace B. Therefore the resistance of trace A will be higher than trace B. Therefore, the I^2R dissipation in trace A will be greater than in trace B and trace A will be hotter.

Conclusion: there is no necessary relationship between current density and temperature.

4.5 Measuring Trace Temperature

There are several different ways we can determine the temperature of a trace. None is without problems, and each has some individual strengths. This section will outline a few of these.

4.5.1 IPC Procedure: One way to measure the trace temperature is to calculate the resistance of the trace at an elevated temperature and compare it to the resistance at a reference temperature (the "change in resistance method.") The IPC outlines this test procedure in its procedure manual [4].

A "standard" test trace is defined as shown in Figure 4.5.

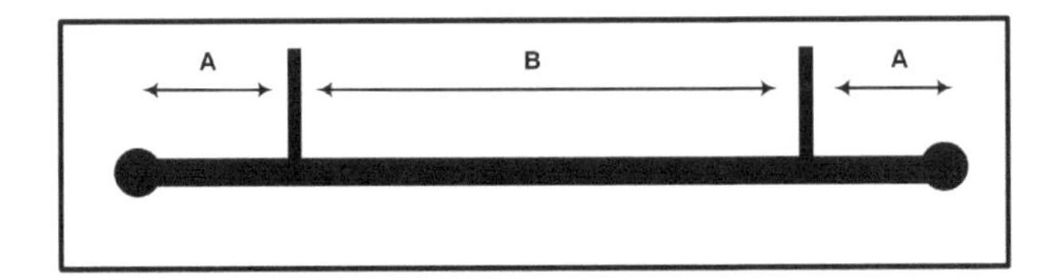

Figure 4.5
IPC "standard" test trace (not to scale.)

The trace length from end-to-end is 12 inches. The pads at each end are 200 mil in diameter. Three inches in from each end (dimension A) is a sense pad, 20 mil wide by 400 mil long. That leaves the active test length (dimension B) at 6". Sense leads, #26 AWG magnet wire, are soldered to the sense pads.

Then the procedure follows these steps:

1. Place the board horizontally in still air.
2. Measure the resistance of the trace at the reference temperature (ambient) by passing a known and constant current through the trace no larger than 100 ma. Use Ohm's Law (R = v/i) to calculate the reference resistance. (Alternatively the resistance could be measure with a precision Ohm meter that is electrically isolated from the trace.)
3. Apply a known (and constant) current to the trace under test.
4. Measure the voltage across the sensing points. Calculate the resistance at the elevated temperature, again by Ohm's Law (or again using a precision Ohm meter.)
5. Calculate the change in temperature using Equation 4.4 (refer back to Section 3.3)

$$\Delta T = \frac{1}{\alpha_0}\left[\frac{R_t}{R_{t_0}} - 1\right]$$

[Eq. 4.4]

Where: ΔT = change in temperature
t_o = reference temperature
R_t = Resistance at the temperature of interest
R_{t0} = Resistance at the reference temperature
α_o = Thermal coefficient of resistivity at the reference temperature.

This procedure measures, by definition, the *average* temperature along the 6" section of the trace. We will show later that there is virtually always a gradient in temperature along a trace, especially near the ends (pads.) On the other hand, the fact that the sense leads are spaced 3" in from each end minimizes the effects of any gradients in the IPC data. So it is important to note that the IPC procedure measures the *average* temperature accurately, but does not account for gradients or hot spots.

4.5.2 Infrared Measurement: Another approach for measuring trace temperature is to use an infrared, or heat sensing microscope or camera (Figure 4.6). This approach measures the temperature at a point along the surface of the trace [5]. The strength of this approach is that it is non-contact and it measures the temperature at a *point* along the trace, which is often what we really want.

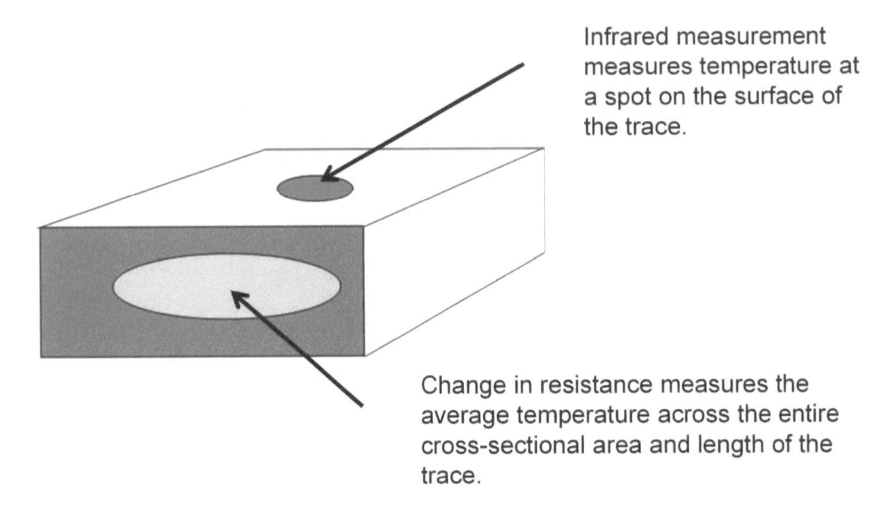

Figure 4.6
Comparison of infrared measurement of trace temperature and approaches that measure the average temperature.

But there are difficulties with this approach, also. For one, it assumes that the "emissivity" of the trace surface is known, which is not always true. Second, it is not as precise as other measures, often with uncertainties measured in several degrees C. On the other hand, this may be the only viable approach for measuring trace temperatures that are very high (as will be covered in Chapter 11). Modern heat sensing measurement tools are getting better and better at overcoming some of the inherent problems with this approach.

4.5.3 Thermocouple Measurement: A third approach to trace temperature measurement is the use of a thermocouple. This is, by definition, a point measurement approach. A thermocouple is placed directly on the trace and the temperature measured directly (Figure 4.7).

Figure 4.7
Use of a thermocouple to measure trace temperature.

This has several advantages. From a data measurement standpoint it is relatively quick and (from a location standpoint) very precise. There are data loggers readily available that can record temperatures as they are measured and automatically graph them. Thermocouples are available with probe points as small as 3 mils and with response times as short as a second or two. However, there is the possibility that the contact with the trace may impact the measurement. One view is that the probe may conduct heat away from the trace and lower the reading. On the other hand, the thermocouple is so small a thermal load, and the copper conductivity of the trace so high, that any loading may be minimal.

Care must be taken that the thermocouple makes good, solid contact with the trace. And the temperature range of many thermocouple probes (because of their coatings) may top out at around 300 to 350°C.

4.6 Temperature Curves

When current is applied to a trace, the temperature increases. But there is a time constant associated with that increase. The pattern of temperature vs. time typically takes one of three general shapes, depending on whether the trace is severely overloaded.

4.6.1 Typical Curve: When current is applied to a trace, the temperature of the trace rises to a point and then stabilizes (Figure 4.8). The time for stabilization (A) is typically anywhere from about 6 minutes to 15 minutes. After that, the temperature remains stable.

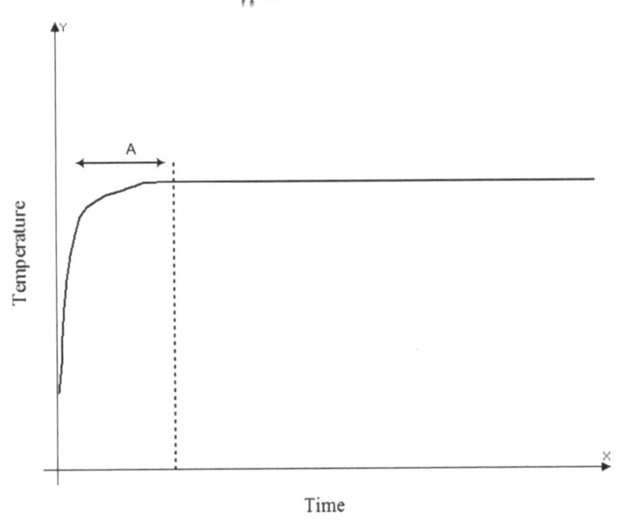

Figure 4.8
Typical temperature relationship.

4.6.2: Heavy Overload: A trace subject to a heavy current overload (so that it melts fairly quickly) follows a pattern typically like that shown in Figure 4.9. The temperature rises in a (more or less) linear way until the trace melts at some point. This is typical of traces that reach a melting temperature within a few (say less than 10) seconds (refer also to Figure 11.1).

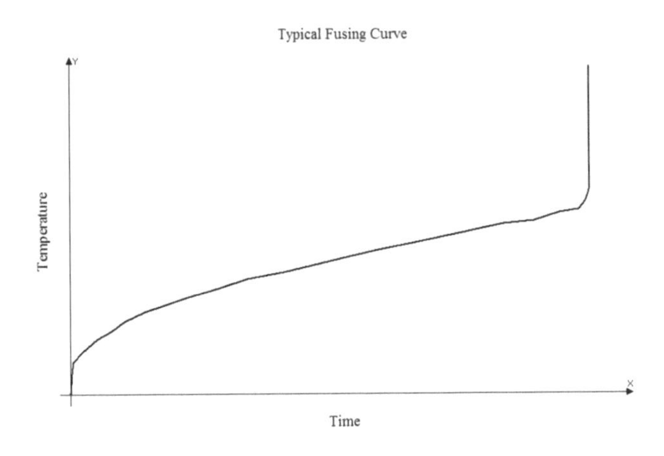

Figure 4.9
Typical fusing curve.

4.6.3 Marginal Overload: Let's say we have a trace that is carrying so much current that it needs supplemental cooling. Cooling may be provided by heat sinks or by forced air. Then let's say that cooling degrades for some reason. Or, let's say the trace is carrying "just enough" current to eventually fuse. This type of failure tends to follow a pattern similar to that shown in Figure 4.10 [6] (refer also to Figure 11.3).

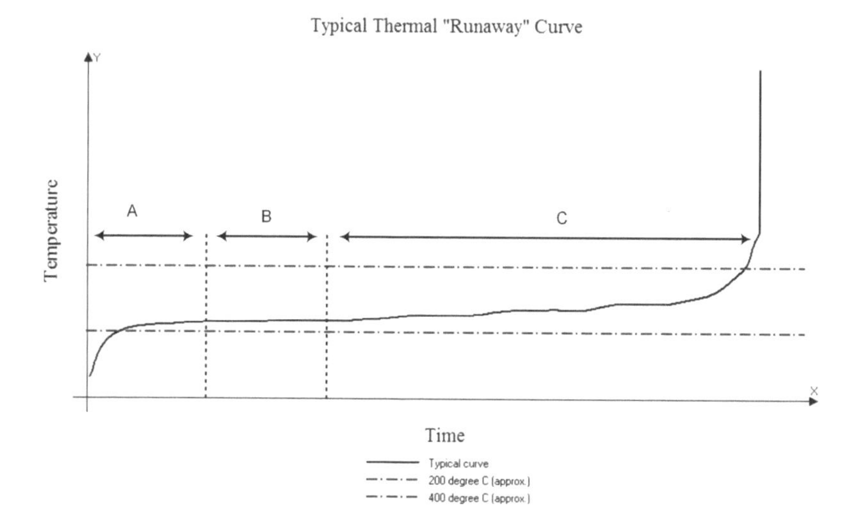

Figure 4.10
Long-term trace failure.

Over the time frame shown as "A" in the figure, the temperature rises to a point and stabilizes. This is exactly like Figure 4.8, above, and happens over the same time frame. Then the temperature seems stable over time frame "B", which can last from minutes to hours. But then the temperature begins to rise. The reason may be the result of several things. Maybe the support cooling approaches began to degrade. Maybe the dielectric began to change (see Section 2.2). But for whatever reason, the temperature begins to increase until the trace ultimately fails (melts) at some spot. This is sometimes described as thermal "runaway."

The temperature range where time range C might start is very situation specific, and will vary from trace to trace. But the range typically is around 220 to 270°C. Once runaway begins, failure is inevitable (unless the current is lowered.) Time frame "C" is typically around 15 minutes or so, but it is very difficult to generalize.

It is important to note that the red lines in the figures represent average temperatures along the trace. These do not represent anything near the melting temperature of copper. The trace will actually melt at a single hot spot along the trace that will be much hotter than the average temperature.

We will cover this topic in much more detail in Chapter 11.

Notes:

1. Thermodynamic engineers make a distinction between the initial temperature (initial condition) and ambient temperature, since the initial condition might, for some reason, be higher or lower than ambient. Electrical engineers are more casual about this and usually (perhaps incorrectly) equate the initial condition with ambient temperature.
2. To the extent that these traces are coated (say conformal or solder coated for example), the cooling may be enhanced or restricted somewhat.
3. See https://en.wikipedia.org/wiki/Heat_transfer_coefficient for an explanation of HTC.
4. See IPC-TM-650 Test Methods Manual, Number 2.5.4.1a, "Conductor Temperature Rise Due to Current Changes in Conductors," freely downloadable at http://www.ipc.org/test-methods.aspx
5. This was the approach used in measuring the temperature in "Printed Circuits and High Currents", Friar, Michael E. and McClurg, Roger H., Design News, Vol. 23, December 6, 1968, pp. 102-107.
6. There is a curve very similar to this in "New Methods of Testing PCB Traces Capacity and Fusing", Norocel Codreanu, Radu Bunea, Paul Svasta, "Politehnica" University of Bucharest, Center for Technological Electronics and Interconnection Techniques, UPB-CETTI, internal research project and report of UPB-CETTI / Winter-2010, (Figure 6). Download at: http://www.cetti.ro/cadence/articles/New_Methods_of_Testing_PCB_Traces_Capacity_and_Fusing.pdf

Part 2

EXPERIMENTAL STUDIES and SIMULATIONS

5 IPC CURVES

IPC 2152, “Standard for Determining Current Carrying Capacity in Printed Board Design,” provides the best data on PCB trace current/temperature relationships that have even been assembled. The publication has over 100 pages and over 100 charts and tables. It covers external traces, internal traces, and traces in a vacuum. And it covers traces from 0.5 Oz. thickness to 3.0 Oz. thickness [1].

5.1: Measuring the Data

The IPC outlines a test procedure that it follows in this type of investigation [2]. A “standard” test trace is defined as shown in Figure 4.5 and reproduced here as Figure 5.1.

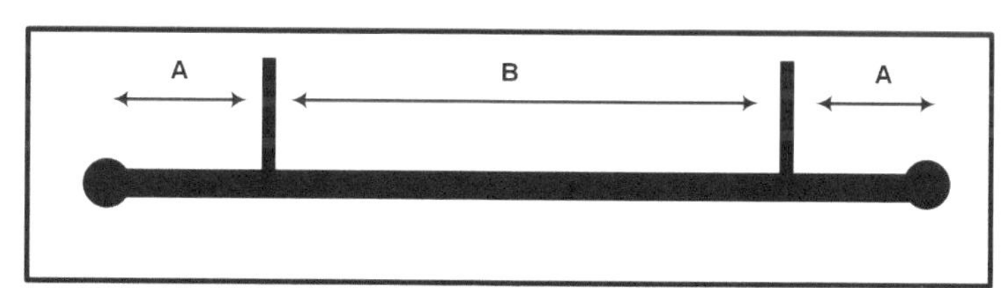

Figure 5.1
IPC “standard” test trace (not to scale.)

The trace length from end-to-end is 12 inches. The pads at each end are 200 mil in diameter. Three inches in from each end (dimension A) is a sense pad, 20 mil wide by 400 mil long. That leaves the active test length (dimension B) at 6”. Sense leads, #26 AWG magnet wire, are soldered to the sense pads.

Although the details in the procedure may seem tedious, they boil down to the following steps:

1. Place the board horizontally in still air.
2. Measure the resistance of the trace across the sensing points at the reference temperature (ambient) by passing a known and constant current

through the trace no larger than 100 ma. Use Ohm's Law (R_{to} = v/i) to calculate the reference resistance.

3. Determine the cross-sectional area of the trace using known trace dimensions or by using Equation 5.1.

[Eq. 5.1]
$$A = \rho * L / R_{to}$$

where: A = cross-sectional area
ρ = Resistivity of the conducting material
L = length of the trace
R_{to} = the measured resistance

4. Apply a known (and constant) current to the trace under test.
5. Measure the voltage across the sensing points. Calculate the resistance (R_t) at the elevated temperature, again by Ohm's Law.
6. Calculate the change in temperature using Equation 5.2 (refer back to Section 3.3)

[Eq. 5.2]
$$\Delta T = \frac{1}{\alpha_0}\left[\frac{R_t}{R_{t_0}} - 1\right]$$

Where: ΔT = change in temperature
t_o = reference temperature
R_t = Resistance at the temperature of interest
R_{t0} = Resistance at the reference temperature
α_o = Thermal coefficient of resistivity at the reference temperature.

7. Repeat step 6 for a variety of currents and voltages and record the data.

There are various strengths and weaknesses with this approach. For one, the IPC approach seems to suggest that ρ, the resistivity, can be taken as published. It does not suggest that it be verified. It also suggests that α_o be calculated "if unknown." but it also seems to suggest that α_o can be determined from published values. As we saw in Chapter 3, both of these values can be suspect and subject to error. (But, to be fair, this comment applies to any trace, anywhere!)

Finally, the procedure requires that the reference resistance (R_{to}) be measured with a "digital multimeter" at a current "not to exceed 100 ma." For a larger trace (say a 2.0 Oz, 100 mil wide trace that is 6" long) this results in an expected resistance of perhaps 0.016 Ohms, resulting in a measured voltage of only 1.6 mv. A voltage that low is tough to measure with precision with a multimeter.

The IPC charts in IPC-2152 plot current vs. cross-sectional area for constant temperature curves. But the data that is taken is for current and temperature for constant cross-sectional area. Thus, data aggregation and interpolation steps (across various test samples) are required before publication. (Again, to be fair, this is a practical and acceptable procedure. It is noted simply to point out that there is an opportunity for error to be introduced in these several steps.)

This procedure measures, by definition, the *average* temperature along the trace. We will show later that there is virtually always a gradient in temperature along a trace, especially near the ends (pads.) On the other hand, the fact that the sense leads are spaced 3" in from each end minimizes the effects of any gradients in the IPC data. So the IPC procedure measures the average temperature accurately, but does not account for gradients or peak temperatures.

5.2: IPC Curves

5.2.1 External results: The results from the testing are provided in curves like that shown in Figure 5.2. The curves plot current as a function of cross-sectional area and create constant-temperature (change) lines on the graph. Each set of curves represent a constant trace thickness.

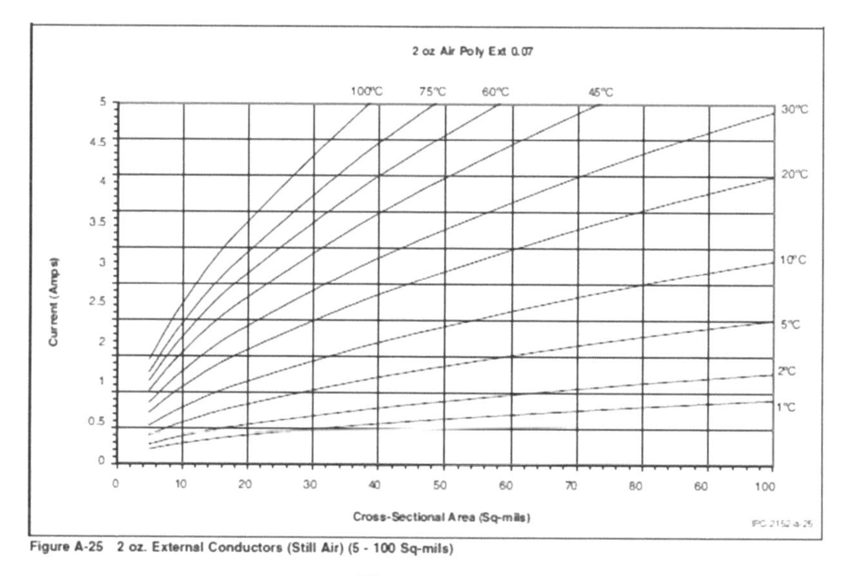

Figure A-25 2 oz. External Conductors (Still Air) (5 - 100 Sq-mils)

Figure 5.2
Typical form of a set of curves in IPC 2152.

It will be useful to modify these curves so that they plot the change in temperature as a function of current and have constant-width curves on the graph. That is, in part, because it is more useful for PCB designers to have data presented this way. It is also because our model (see Equation 4.3) is expressed this way. Finally, it is also because thermal simulations tend to be formulated this way. So we need to transpose the IPC data to a different format.

A convenient and efficient way to do this is through the use of a digitizing program [3]. When we do that, we obtain a set of graphs that look like Figure 5.3.

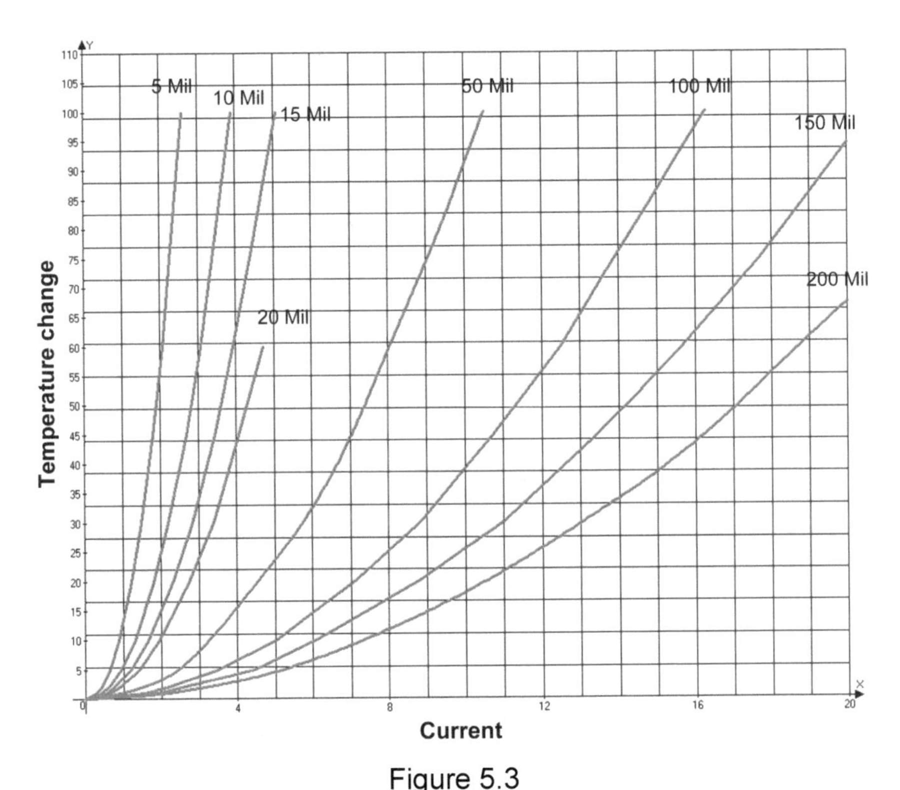

Figure 5.3
The 2 Oz. external IPC curves re-drawn to a different set of axes.

5.2.2 External IPC Data Equations: The next task is to try to fit these curves with an equation. One way to do this is with a multiple regression analysis using a resource such as any current spreadsheet. Another way is to start with our model (Equation 4.3, repeated here as Equation 5.3):

$$\Delta T \propto \frac{I^2 R}{(w + Th)}$$

[Eq. 5.3]

Recognize that Resistance, R, is inversely proportional to area (width*thickness), so this model can also be expressed as Equation 5.4:

[Eq. 5.4]
$$\Delta T \propto \frac{C^2}{Width^{a1} * Th^{a2}}$$

Where: ΔT = Change in temperature
C = current
Width = trace width
Th = trace thickness
a1 and a2 are undetermined constants that are approximately 1.0 to 1.5

After a little work it was determined that the best fit for these curves was an equation of the form:

[Eq. 5.5] $\Delta T = 215.3 * C^2 * W^{-1.15} * Th^{-1.0}$

Plotting this equation onto Figure 5.3 (with the appropriate values for W and Th) gives the results shown in Figure 5.4 (solid black lines are from the IPC data, the dotted red lines are the equations.)

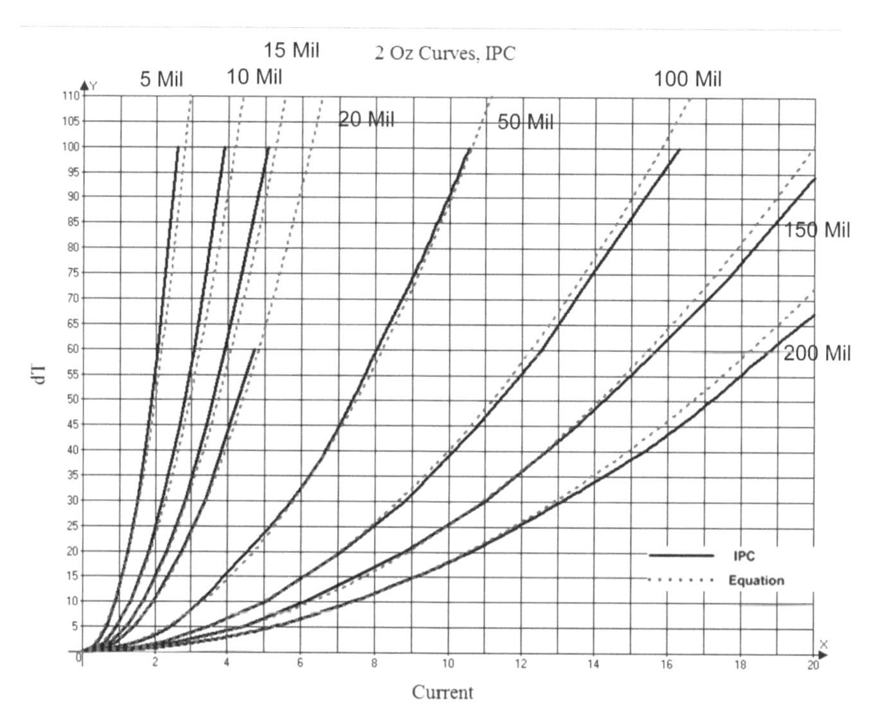

Figure 5.4
2 Oz. external IPC curves of Figure 5.3 fitted with Equation 5.5

This result is impressive, but what would be even more meaningful would be if this same equation, for 2 Oz. external data, would also fit the 3 Oz. external IPC data. If that were the case, we would have a fair amount of confidence in both the IPC data and the equations.

Figure 5.5 illustrates similar results for the 3 Oz. external IPC data. The curves were generated in the same fashion as the 2 Oz. curves were, and the equation is the same, differing only in terms of the appropriate variables (width and thickness).

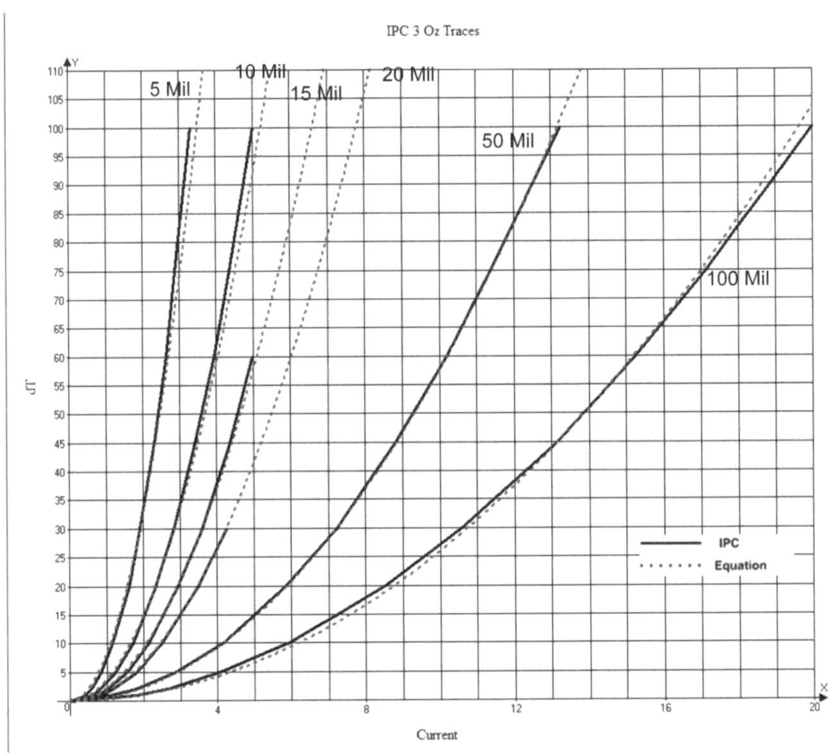

Figure 5.5
3 Oz. external IPC curves fit with Equation 5.5

The fits are obviously extremely good. This gives us very high confidence in the equations.

We can't do a similar fit test for 1 Oz. traces because there is no separate 1 Oz. external data in IPC 2152. Section 5 in IPC 2152 provides a set of charts for overall use, but they reflect a consolidation of all data, not exclusively 1 Oz. data. Similarly, there are no 1 Oz. external data charts in the Appendix. We have produced Figure 5.6, however, based on Equation 5.5. In the next chapter we will fit this curve with thermal simulation results for comparison with the 2 Oz. and 3 Oz. curves.

As an aside, it is worth noting that Equation 5.5 also proves that the change in temperature is not a function of current density alone, as some people might think (see Section 4.4). Since the coefficients for the width and thickness terms are different, the form factor of the trace (i.e. whether it is wider or thicker for the same cross-sectional area) matters. Although the coefficients are not exactly what we expect from our initial model (for reasons that are not exactly clear), the closeness of the fit (and the fit of the thermal models in the next section) cannot be ignored.

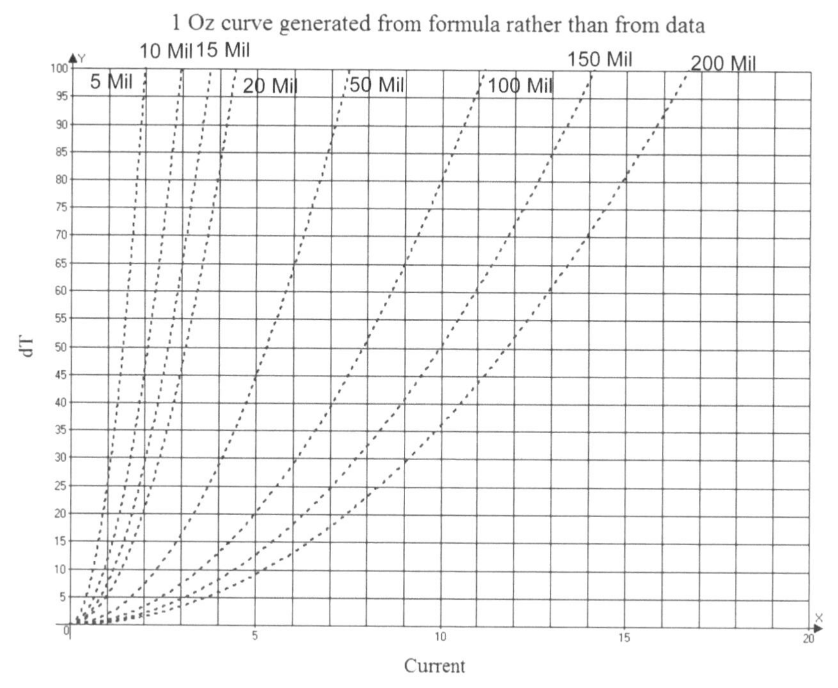

Figure 5.6
1 Oz. curves generated from Equation 5.5

5.2.3 Internal IPC Data Equations: A real surprise from the IPC 2512 study and data was that the internal races actually cool *better than* the external traces do. The internal traces are a little cooler than are the external traces.

Fitting the Internal IPC data does not go quite so smoothly. To save space in this paper we have put the four internal fitted curves in Appendix 3. On the surface, the curves and the fits look excellent. The problem is, in this case, the equations are not all exactly equal. They are quite close, but not exactly equal. The primary difference is in the constant term. Table 5.1 (below) provides the coefficients for the equations.

We will look at how the thermal models fit the internal data later on. At this point, my suspicion is that the differences reflect a control issue in the study.

As we will see in Chapter 6 the temperature of a trace is very sensitive to a variety of factors, only some of which are easily controlled. The differences here between trace widths are not nearly as great, however, as some of the other differences we will explore in Chapter 6.

5.2.4 IPC Vacuum Data: Traces are cooled by three effects: (a) heat conducting away from the trace through the board, (b) heat convected away from the board into the surrounding air, (c) heat radiating away from the board into "space." In a vacuum, there is no surrounding air. So heat that gets to the surface (either because the trace is on the surface or because the heat has conducted through the board to the surface) can only radiate away. This is much less efficient than convection through the air, so traces in a vacuum understandably run quite a bit hotter than otherwise. And it appears not to make too much difference whether we are talking about internal or external traces (IPC 2152 does not provide separate data for internal and external traces in a vacuum. Nor does IPC 2152 provide any 1 Oz. data for traces in a vacuum.)

The curves for traces in a vacuum (and their fitted equations) are provided in Appendix 3. Figure 5.7 illustrates the difference between some 2 Oz. external, internal and vacuum traces. The pattern is similar for all other sizes of trace.

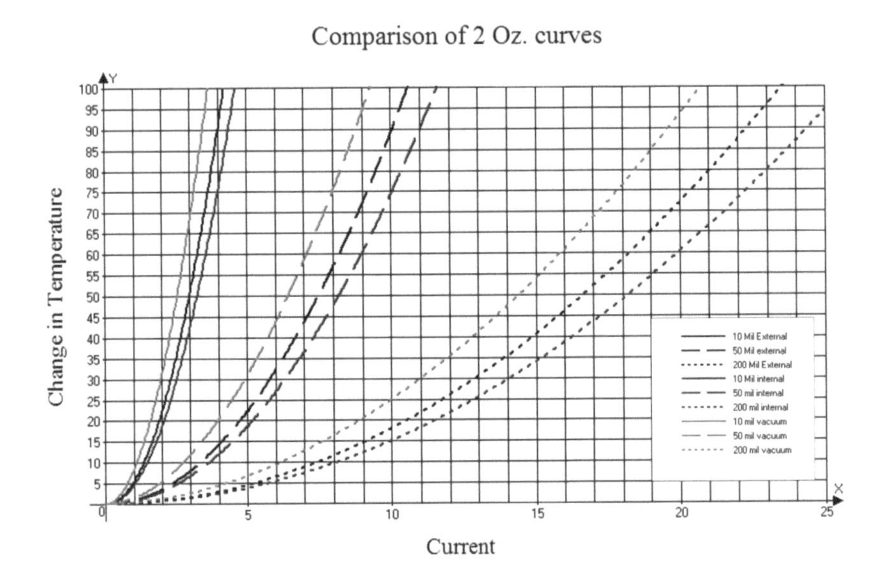

Figure 5.7
Comparison of 2 Oz. external, internal and vacuum traces of width 10 mil, 50 mil, and 200 mil.

Table 5.1 shows the approximate equations used for fitting the curves. The complete set of equations is provided in Appendix 4.

Data	Constant	C^	W^	Th^
External				
All	215.3	2	-1.15	-1.0
Internal				
0.5 Oz.	110-130	2	-1.10	-1.52
1 Oz.	200	1.9	-1.10	-1.52
2 Oz.	300	2	-1.15	-1.52
3 Oz.	225-300	1.9	-1.15	-1.52
Vacuum				
0.5 Oz.	210-235	1.9	-1.10	1.52
2 Oz.	480	1.9	-1.10	1.52
3 Oz.	460	1.95	-1.15	1.52

Table 5.1
Coefficients for all the IPC equations

Notes:

1. Copper trace thickness is typically measured in ounces of copper. A one-ounce trace is approximately 1.3 mils thick.

2. See IPC-TM-650 Test Methods Manual, Number 2.5.4.1a, "Conductor Temperature Rise Due to Current Changes in Conductors," freely downloadable at http://www.ipc.org/test-methods.aspx

3. I used a program called GetData Graph Digitizer, available here: http://www.getdata-graph-digitizer.com/ . See also Douglas Brooks and Johannes Adam, "Trace Currents and Temperatures Revisited," available at www.ultracad.com.

6 THERMAL SIMULATIONS

In the previous chapter we looked at the IPC curves and fit them with equations. In this chapter, we will independently try to verify the curves using a computer thermal simulation program. If we are successful at doing that, then we will simulate other typical conditions not covered by the IPC curves, things like adjacent trace and underlying planes, and see what impact those things might have on the results.

6.1 Modeling Traces:

Dr. Johannes Adam was kind enough to provide a copy of ADAM Research's TRM 1.8.10 simulation software for that purpose. TRM (Thermal Risk Management) was originally conceived and designed to analyze temperatures across a circuit board, taking into consideration the complete trace layout with optional Joule heating as well as various components and their own contributions to heat generation. Although the program could be adapted to the measurement of an individual trace, it was not originally conceived with that use in mind. Consequently, a couple of adjustments and adaptations had to be made.

For those of us in the printed circuit board industry, a thermal simulation model is to trace temperature calculations what a field effect model is to trace impedance calculations. And the approach is not too different. The 3-D structure is analyzed by first looking at a tiny cube [1] within the structure (see Figure 6.1.) At this micro level, the computations are relatively straightforward to define. So you do the calculation for one cube, then the next cube, and then the next one --- etc. And you keep going in this iterative process until you have solved for the entire structure.

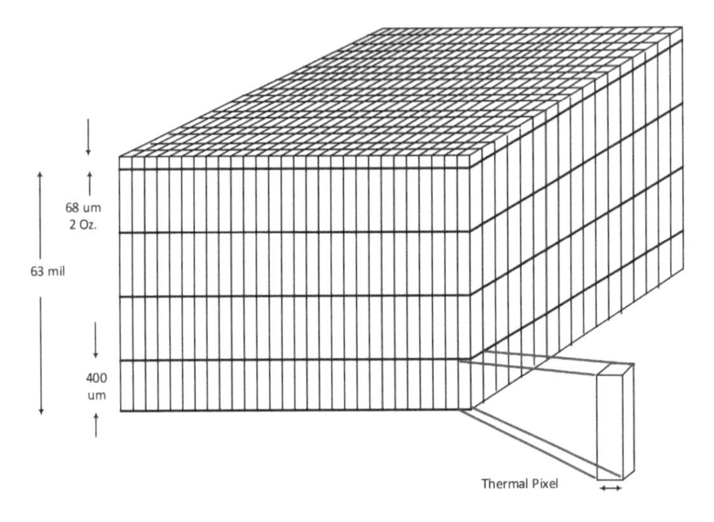

Figure 6.1
A model consists of a large number of square or rectangular "cubes."

Figure 6.1 represents a very small portion of the model we will develop in Section 6.2, below. There will be a very large number of "cubes" in this analysis, and a great many calculations for each. The analysis becomes a very large matrix algebra computational problem, very difficult for an individual but perfect for a desktop computer. Any practical problem can require a large number of resources. A reasonable problem, with reasonable accuracy, can require a few minutes to several hours to solve, even with a powerful 64-bit desktop computer.

I will illustrate a couple of the adjustments we had to make below, as I describe one of the models. But two comments deserve mention up front. First, as mentioned in Section 3.3, resistivity, and therefore resistance, increases with temperature. So if we initiate an analysis with a certain resistivity in mind, and reach an elevated temperature, the resistance will have changed at that elevated temperature. That means the trace will be hotter than the analysis calculates. TRM runs in two selectable modes. One goes directly to a solution and one takes changing resistivity into account.

In the second mode, the program executes with initial values and reaches a solution. Then the program recalculates the resistance of the trace at the new temperature and executes (loops) a second time. And it keeps doing this. The user sets the number of loops as part of the setup. Experimentally we determined that two loops were sufficient to reach (acceptable) stability at lower temperatures and for internal traces and 4 loops (or more) were needed at higher temperatures.

Secondly, setting up a model requires defining something called the "Heat Transfer Coefficient" (HTC). This is a coefficient that defines how effectively a surface transfers heat to another medium. In this situation, it refers to how effectively the board and trace transfer heat to the surrounding air. This coefficient, found in the literature with letter h, is often not familiar to electric engineers (other than perhaps the thermal resistance Rth). However, only h provides the necessary coupling between temperature inside the board and the ambient (otherwise heat could not leave a board). In practice HTC contains the contributions of convection plus radiation. While HTC can be estimated quite well for flat and uniformly heated plates, it is not always intuitive what the value should be for a single trace on a board. But more importantly, HTC increases with temperature, and it is not at all clear how it increases for an individual trace. Here is how Johannes and I attacked this problem:

We set up a model and entered a reasonable value for HTC. We solved the model and compared the result to the equations established above. We found the base value (i.e. the value of HTC for low currents and temperatures) pretty quickly. It was 10 W/m^{2}K [2]. Then we solved the models at higher currents and determined what values of HTC were required to fit the equations. Higher values were expected and determined.

At lower currents (temperatures) the model results were pretty insensitive to HTC, often differing by 1 degree C or less. At higher currents (temperatures) a range of HTC from 11 to 14 only resulted in a temperature difference of around 10 degrees (in one case 20 degrees.) So the determination of the correct value for HTC was more like "tweaking" the results of the model.

The range of HTC's needed for external and internal traces is shown in Figures 6.2 (a) and (b). The range falls between the red lines.

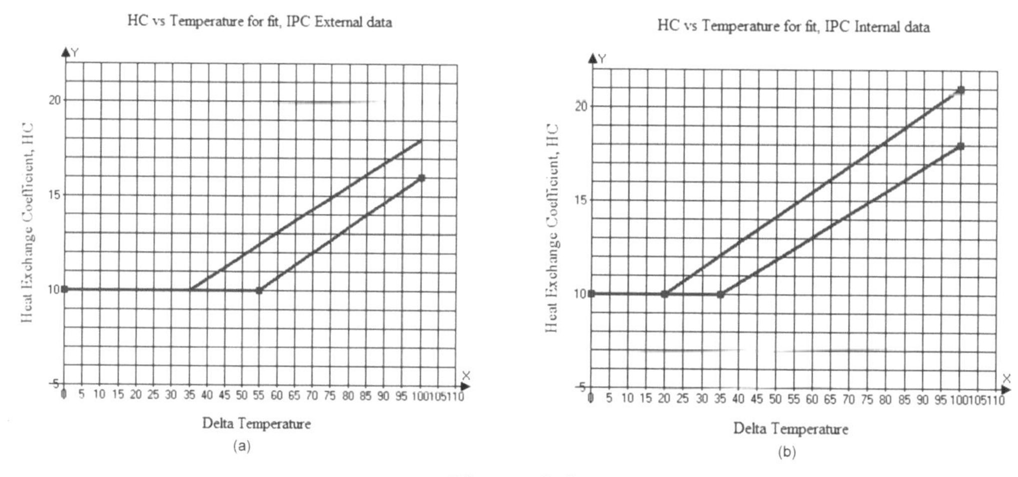

Figure 6.2
Approximate range of HTC W/m^{2}K for (a) external and (b) internal traces.

While it may seem like were somehow forcing the model to fit the data, think of the initial steps as more like *calibrating* the model. Once we knew what the predictable range was, we could then pre-set the HTC value and get meaningful results.

6.2 The IPC Modeling Process:

It is not my intent in this section to teach someone how to run a TRM thermal simulation model. I will go through the steps to illustrate what is involved, but I won't cover every single step in the process. Furthermore, this illustrates one software's approach. There are other software programs out there that can perform similar simulations. I am using TRM as an illustration here because that is what we used, and that is what I know.

The IPC 2 Oz., 200 mil trace is modeled as shown at the top of Figure 6.6 (an actual image from the output folder.) The modeled trace is approximately 300 mm (11.8 in) long and 5 mm (200 mil) wide. The sensor pads are place approximately 6.0" apart. There are 4.0 mm wide traces attached to the ends of the sensor lines to simulate the leads that are soldered onto the IPC test fixture.

Figure 6.3 shows the "Build Board/Prepare" menu for setting up the board. The board outline is 350 x 45 mm. The default values for the conductor and dielectric are specified (here normal copper and Polyimide). On the right is shown a copy of the material database for Polyimide. The arrows highlight the entries for the in-plane and through-plane entries for the thermal conductivity coefficient (these entries are editable.) The "Thermal pixel" is specified as 0.2 mm. Think of the thermal pixel as the minimum length of the side of one of the cubes shown in Figure 6.1. This must be smaller than any other dimension in the x,y plane. The smaller the cube, the more cubes there are in the analysis and the larger are the computational matrices. A thermal pixel of 0.2 is not particularly small, and this analysis will run fairly quickly. If we were analyzing a via model, the thermal pixel might be specified as 0.02 (or maybe even smaller) and the analysis would take much longer.

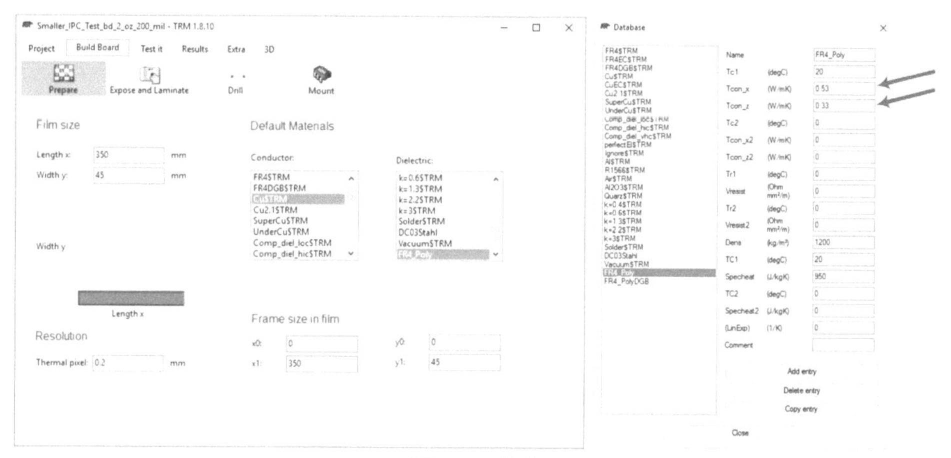

Figure 6.3
The project prepare menu

Next is the "Expose and Laminate" menu, Figure 6.4. Think of this as the board stackup. Of note, the top layer (which will be copper foil) is 68 µm (2 Oz.). The board itself is 1600 mm (63 mils) thick. It is broken into four separate layers of 400 mm each for technical reasons. (Figure 6.1 models this stackup.) This provides a little more precision to the model. The conductor and dielectric specifications set here over-ride the default values set in Figure 6.3.

Smaller_IPC_Test_bd_2_oz_200_mil - TRM 1.8.10

Project | Build Board | Test it | Results | Extra | 3D

Prepare | Expose and Laminate | Drill | Mount

Level	Name	Type	File	View	FR4 white?	Thick (mu)	Conductor	Dielectric	Expose
1	Top	pre		View	☑	68	CuSTRM	FR4_Poly	Expose
2	Core	pre		View	☑	400	CuSTRM	FR4_Poly	Expose
3	Core	pre		View	☑	400	CuSTRM	FR4_Poly	Expose
4	Core	pre		View	☑	400	CuSTRM	FR4_Poly	Expose
5	Core	pre		View	☑	400	CuSTRM	FR4_Poly	Expose
					☐				

Figure 6.4
Expose and laminate menu.

There are separate menus for specifying the conductor and dielectric values in very precise detail. Of importance to us would be the resistivity and thermal coefficient of resistivity values for the conductor, and the thermal conductivity values for the dielectric. Since we don't know the thermal conductivity of the specific type of Polyimide used in the IPC studies, our model uses the default values for Polyimide, 0.53 for both in-plane and through-plane conductivities.

Next is the Environment menu, Figure 6.5. This is pretty straight forward. The ambient temperature is set at 20° C and the HTC coefficient (discussed above) set as 11 W/m^{2}K.

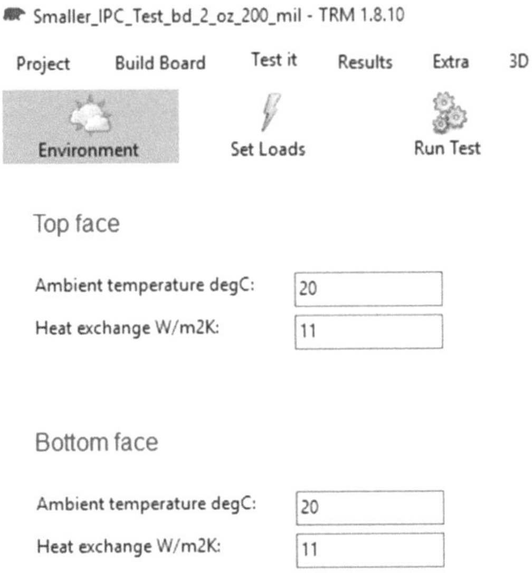

Smaller_IPC_Test_bd_2_oz_200_mil - TRM 1.8.10

Project Build Board Test it Results Extra 3D

Environment Set Loads Run Test

Top face

Ambient temperature degC: 20

Heat exchange W/m2K: 11

Bottom face

Ambient temperature degC: 20

Heat exchange W/m2K: 11

Figure 6.5
Environment menu.

Then we come to the "Set Loads" menu, Figure 6.6. This is the heart of the set-up. This is where we define the traces, pads, test points, and anything else we want to specify. Remember that the board is 350 mm wide and 45 mm high. Think of the lower, left hand corner of the board as having coordinates 0,0. The power and return pads start at coordinates 30,15 and 330,15, respectively. They are each 5 mm wide and 5 mm high. 16 Amps enters the Power pad (+16) and exits the Return pad (-16). The trace starts at 33.2,15 and is 300 mm long and 5 mm high. Since the trace overlaps the pads, the trace length distance between the pads is 295 mm.

The figure above the Set Loads menu is a copy of the trace being modeled. The numbers on this figure correspond to the lines in the Set Loads menu.

The PadL and PadR traces are the sense lines. They start at 105,17 and 255,17, respectively. They are 0.6 mm wide and extend upwards 20 mm. They are 150 mm apart, almost 6 inches. The Trace_Left and Trace_Right traces simulate the magnet wires soldered to the IPC test future. Finally, I have added a thermocouple (TC_Middle) at the midpoint of the trace. This will report the model temperature at that point in the final model results. (Later we can add multiple thermocouples to a model to track any thermal gradients that might exist.)

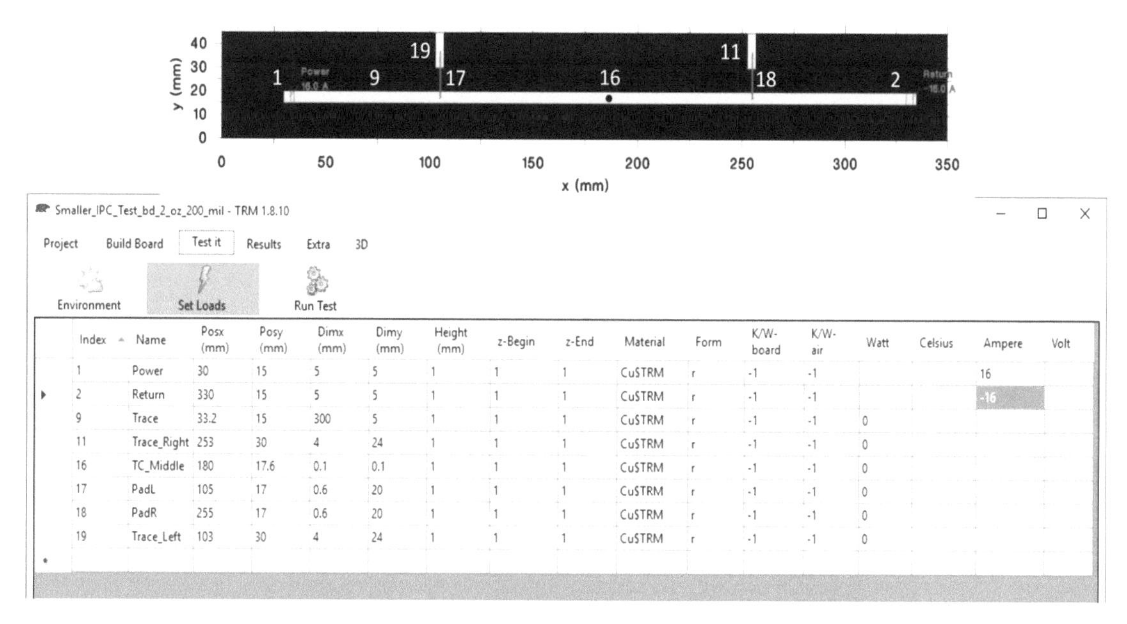

Index	Name	Posx (mm)	Posy (mm)	Dimx (mm)	Dimy (mm)	Height (mm)	z-Begin	z-End	Material	Form	K/W-board	K/W-air	Watt	Celsius	Ampere	Volt
1	Power	30	15	5	5	1	1	1	CuSTRM	r	-1	-1			16	
2	Return	330	15	5	5	1	1	1	CuSTRM	r	-1	-1			-16	
9	Trace	33.2	15	300	5	1	1	1	CuSTRM	r	-1	-1	0			
11	Trace_Right	253	30	4	24	1	1	1	CuSTRM	r	-1	-1	0			
16	TC_Middle	180	17.6	0.1	0.1	1	1	1	CuSTRM	r	-1	-1	0			
17	PadL	105	17	0.6	20	1	1	1	CuSTRM	r	-1	-1	0			
18	PadR	255	17	0.6	20	1	1	1	CuSTRM	r	-1	-1	0			
19	Trace_Left	103	30	4	24	1	1	1	CuSTRM	r	-1	-1	0			

Figure 6.6
The "Set Loads" menu, the heart of the model.

Finally, Figure 6.7 shows the "Run Test" menu. Most of the entries here are details, but one entry of note is the "Temperature dependent" box is checked. This tells the model to run several loops. After each loop the resistivity of the conductor is recalculated for the new temperature and the model runs again. Thus, this is an iterative process. The model gets closer to a stable result with each loop. Experience shows that 3 loops is sufficient at lower end temperatures (40° to 50° C) and 4 loops is sufficient for temperature up to about 100° C.

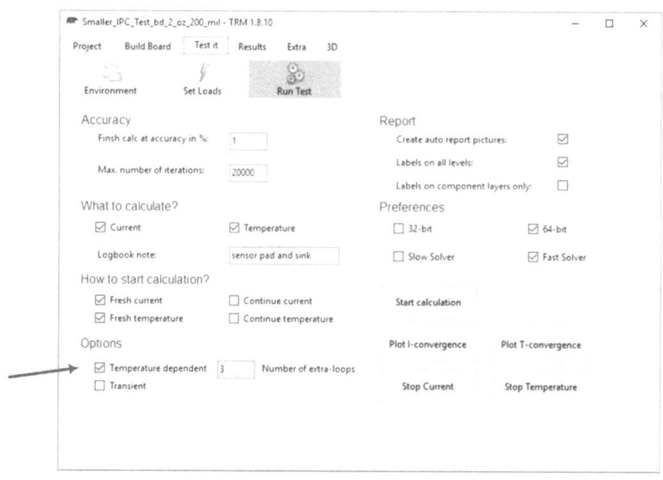

Figure 6.7
The Run Test" menu.

After the model runs, a large number of files are generated. In this case the total number of files was 120! A copy of the output folder is shown in Figure 6.8. Four entries are circled. The file report.txt is a text file that summarizes all the setup and simulation data (ADAM-Research can reconstruct the simulation model from this one file alone.) It contains a large amount of data, including the final temperature at the thermocouple --- the number we want. That temperature is 65.5° C, which represents a change in temperature of 45.5° from the ambient. The circled file "tcon01.png" is the image at the top of Figure 6.6. The circled file "temp01.png" is Figure 6.9, below. The folder named "fields" is filled with another large set of data files which are used for subsequent post processing by the TRM software.

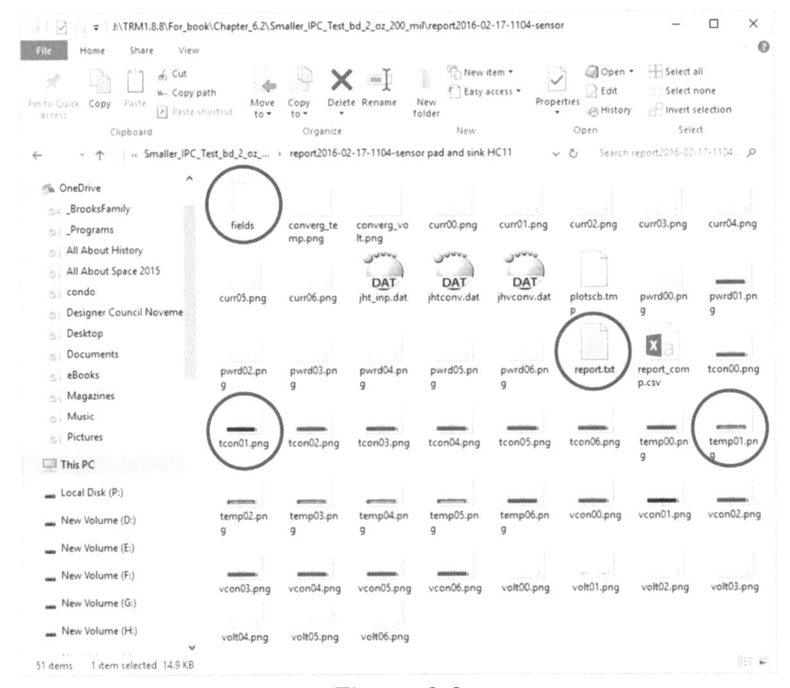

Figure 6.8
The output folder for our model simulation.

Figure 6.9 is the thermal image file (temp01.png) for the simulation.

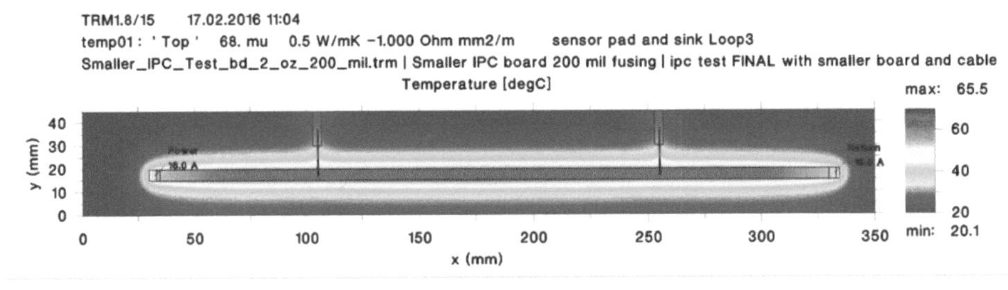

Figure 6.9
Thermal image of the IPC test board.

Several other simulations were run on the 2.0 Oz. IPC model. Their results are shown in Figure 6.10. The result of the specific simulation described here is circled in red. Simulations were also run on 3.0 Oz. external traces, shown in

Figure 6.11. Since there is no pure IPC chart for external 1.0 Oz. traces, Figure 6.12 plots the simulation results only against Equation 5.5. Running a simulation for every point on every curve is not practical, but the wide range of these results should provide confidence that the equation and the thermal simulations are valid.

One point in particular should be emphasized. ***In all three sets of external curves, there is one single equation (Equation 5.5) and one single TRM model (differing only in trace width, trace thickness, and HTC).***

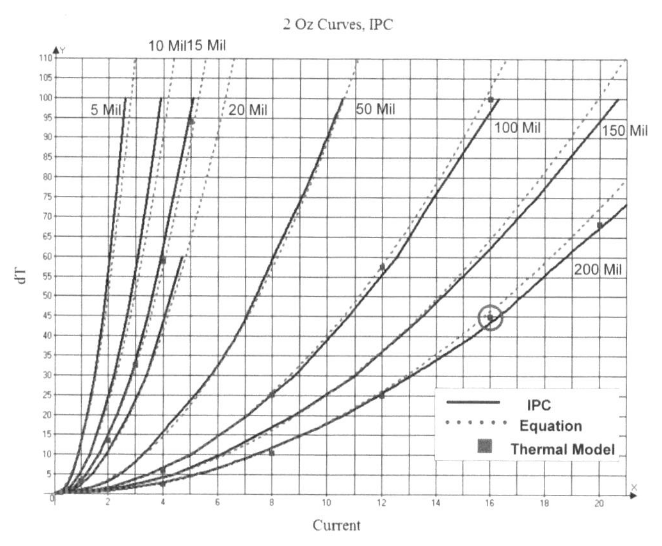

Figure 6.10
IPC 2 Oz. external curves including results of thermal simulation models.

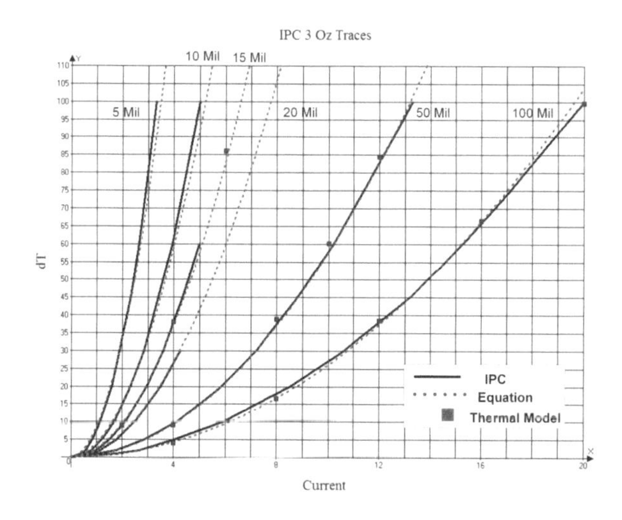

Figure 6.11
IPC 3 Oz. external curves including results of thermal simulation models

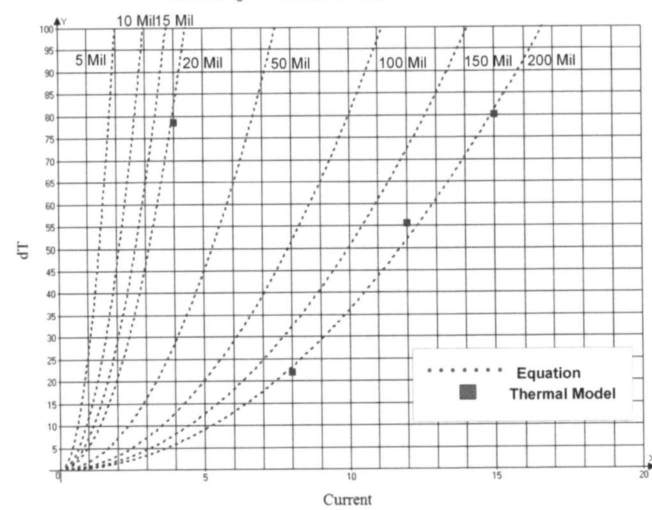

Figure 6.12
1.0 Oz. IPC external curves derived from Equation 5.5 with thermal simulation results

Simulation models were run on selected Internal and Vacuum IPC traces. The results from those simulations are provided in Appendix 5.

There is one item of particular note regarding traces in a vacuum. As mentioned above, the two HTC components are convection and radiation. Very roughly speaking, they are approximately equal in a situation like this. For simulations of traces in a vacuum, the HTC values were in the range of 5 to 9, approximately half that for simulations of traces in air. This is entirely consistent with expectations.

6.3 Sensitivities: Layout Parameters

A simulation tool such as TRM allows us to simulate much more varied and complex trace structures than we can, from a practical standpoint, measure in a laboratory environment. In this section we will look at a variety of parameters and see how the thermal results might change as we change those parameters. Thanks to the generosity of Prototron Circuits (see Acknowledgement), I did have some test boards to actually evaluate and measure a few of the parameter changes.

6.3.1 Experimental Board Design: For some of the analyses to follow, I had access to a test board to check measurements experimentally against a simulation model. So I put together a "standard" simulation model for analysis. Figure 6.13 illustrates a diagram of that model.

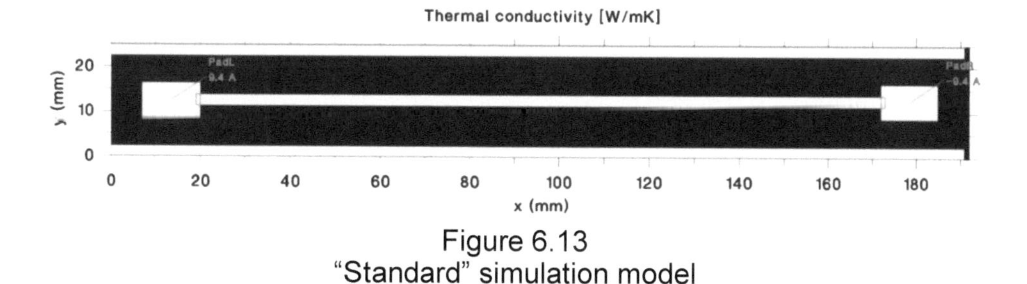

Figure 6.13
"Standard" simulation model

The board outline is 192 mm x 25 mm. There are copper strips along the top and bottom edges to simulate the adjacent traces on the test board. The pads (left and right) are 0.5 mil x 0.3 mil (13 x 7.8 mm). The trace length is 6 inches (152 mm). The trace itself is centered on the centerline of the board. Its width is adjusted as necessary for the particular simulation. A second copper strip is added, if appropriate, (not shown) to simulate a parallel trace. Trace thickness is 1.9 mils (2.7 mils for the 27 mil wide trace). The board thickness is 63 mils (1600 µm). Board thermal conductivity was measured by C-Therm Technologies (see Acknowledgement and Appendix 1) as 0.68 (in-plane) and 0.51 (through-plane). This much represents actual traces I had available for experimental testing.

This simulation model was then modified as necessary for additional simulations (different trace lengths, additional planes, etc.)

6.3.2 Small Trace Widths: The current/temperature curves are very steep for narrow traces. Figure 6.14 illustrates the curves for 1 Oz., 5 and 10 mil wide internal and external traces. They are based on the equations from Table 5.1.

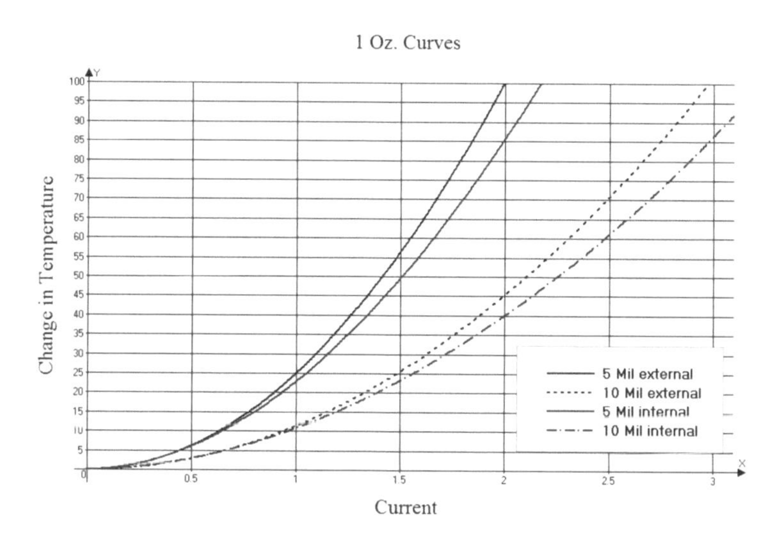

Figure 6.14
1 Oz., 5 mil and 10 mil external and internal curves

For 5 mil traces the difference between a 500 Ma current and a 1.0 Amp current is about 20 degrees C. A 1.0 Amp current has about a 25 degree temperature increase, while a 2.0 amp current has an almost 100 degree increase! The situation is only slightly better for 10 mil wide traces. As soon as we start carrying any significant currents at all, the trace temperatures increase quickly.

Figure 6.15 shows the 5 mil and 10 mil wide external traces, compared to 4 mil and 9 mil traces, respectively. Depending on how great the fabrication tolerances are, trace temperatures unexpectedly high might develop.

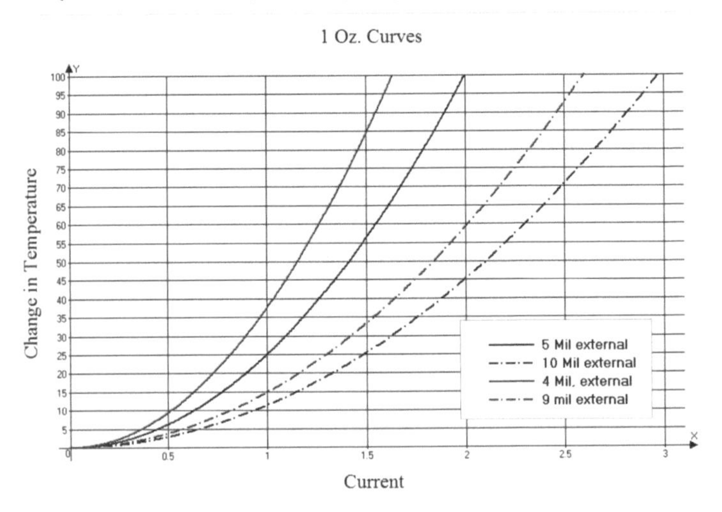

Figure 6.15
Comparing possible fabrication tolerances for 5 mil and 10 mil external traces.

The bottom line is that, if trace temperature for very narrow traces is a concern, designers might want to be particularly conservative in specifying trace widths.

6.3.3 Changing Trace Length: Our standard Test trace length is 6.0 inches. It is interesting to see how sensitive the temperature is to trace length. I ran test lengths of 6.0 inches, 4.0 inches, 2 inches, and 1.0 inch, as well as 2.0 inches and 1.0 inch with a heat sink under the pads (a simulated plane on the opposite side of the board. The plane did *not* extend under the trace.) The results are summarized in Table 6.1.

There is a clear tendency for the temperature to drop as the trace shortens, but it is not a particularly strong tendency unless we are dealing with very short traces. The size of the pads has some influence on the degree of temperature drop.

Test condition	Center Temperature
6 Inch trace Measured	67.0
6 inch trace, simulated	69.2
4 inch trace, simulated	68.0
2 inch trace, simulated	60.9
2 inch trace, simulated, plane underneath pads	59.3
1 inch trace, simulated	49.0
1 inch trace, simulated, plane underneath pads	46.3

Table 6.1
Trace temperature as a function of trace length.

6.3.4 Thermal Gradient: There is a clear thermal gradient from the center of the trace (usually the highest temperature point) out to the edge of the trace. A typical thermal gradient is illustrated in Figure 6.16. This is for a 100 mil wide, 6.0 inch trace.

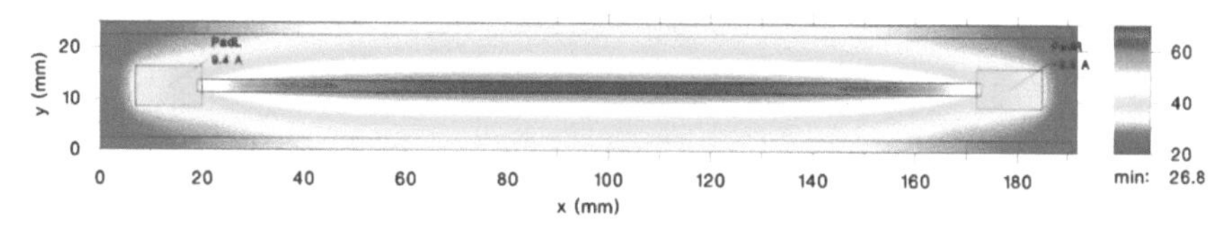

Figure 6.16
Thermal gradient for a 6.0" 100 mil wide trace carrying 9.35 Amps.

The measured and actual thermal gradients for the 200 mil, 100 mil and 27 mil wide traces are graphed in Figure 6.17. Notice how sharply the thermal gradients drop off as you get closer to the edge of the trace.

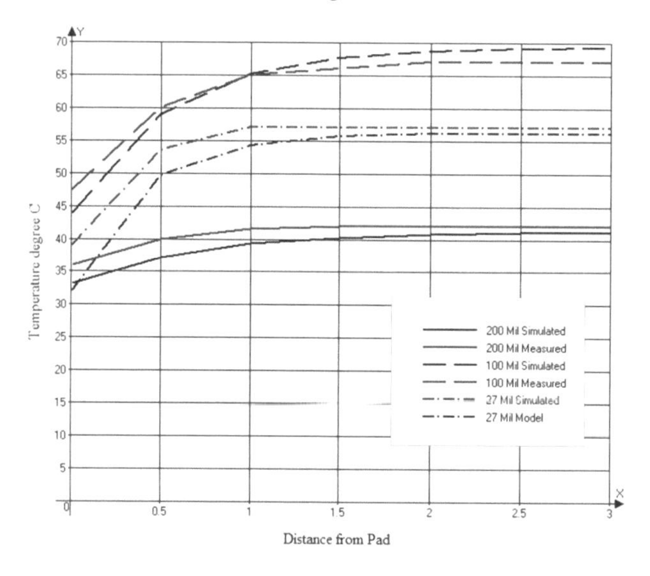

Figure 6.17
Measured and actual gradients for three trace sizes.

6.3.5 Transient Response: Traces do not heat immediately. They take some time to reach temperature, sometimes a surprisingly long time. This can be a problem for people running laboratory tests of trace current/temperature relationships; they must be sure they are waiting long enough for the trace temperatures to stabilize.

Using a simulation model for a 1.0 Oz., 6" long, 200 mil wide external trace driven at 15 Amps, Figure 6.18 shows a curve of how long it takes the trace to reach a stabilized temperature of 94.6 degrees C. It takes almost 3.5 minutes for the trace to get within 90% of its final value, almost 5 minutes to get within 95% of its final value. A *normalized* curve for an internal trace rising to the same temperature is also shown as the red dotted curve. The internal trace rises slightly more slowly than does the external trace because internal traces cool more efficiently than do external traces.

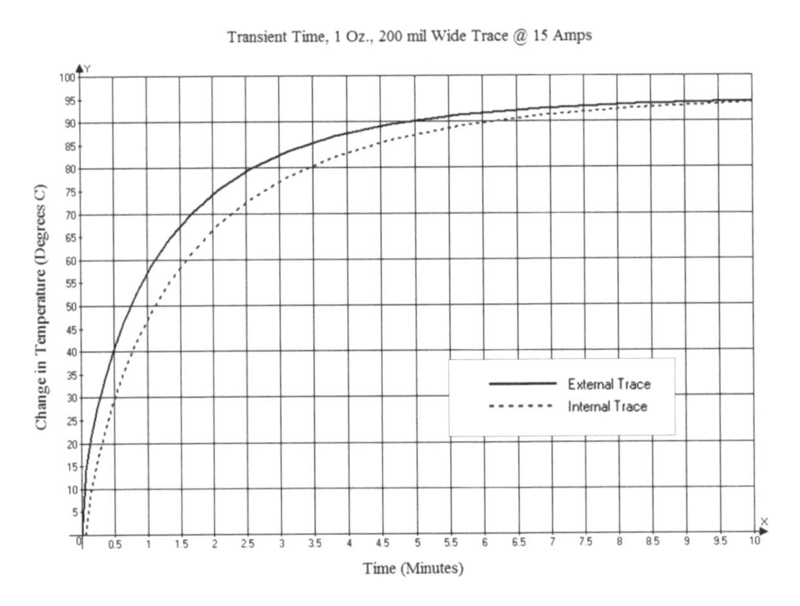

Figure 6.18
Heating time for a 1 Oz., 200 mil wide trace carrying 15 Amps.

6.3.6 Presence of Planes: Most of today's boards have planes on them. The planes play a significant role in cooling the traces above them, because the planes offer a significant heat conducting path away from the trace. Earlier we looked at a model of a 100 mil wide trace with no plane. In Figure 6.19 we see what happens when we add a plane to the board; first to the bottom layer of the board, and then 10 mils underneath the trace layer. The temperature ranges from 69.2° C to 54.1° C to 45.9° C, respectively, for the three cases.

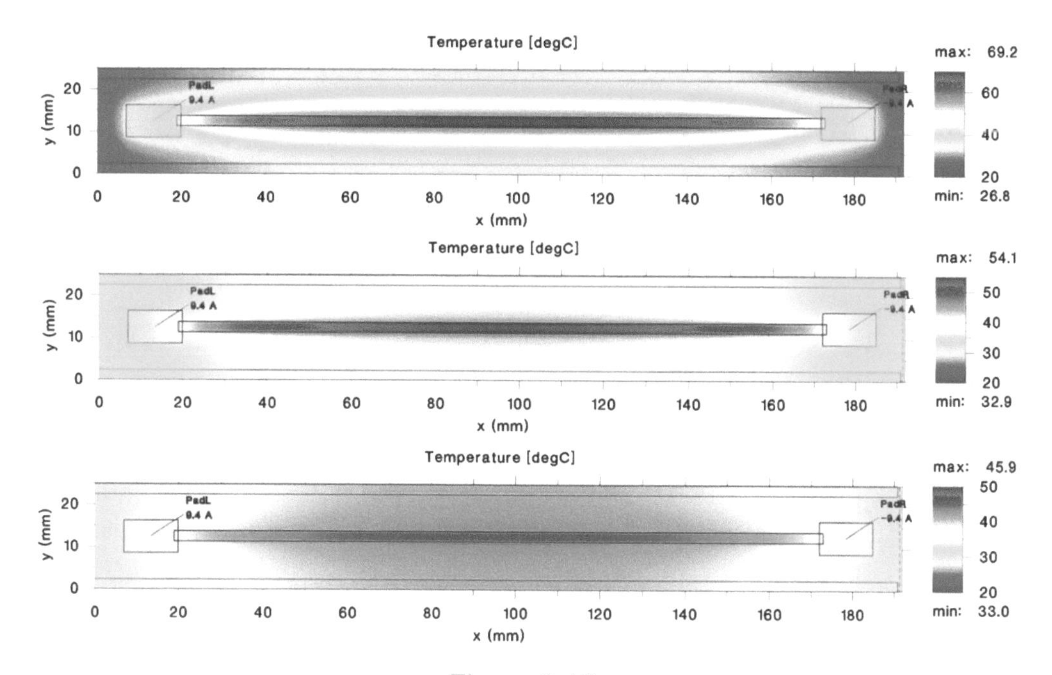

Figure 6.19
Impact of planes: no plane (top), plane on bottom layer (middle), plane 10 mils under traces (bottom).

But even more importantly, look at the thermal "halo" around the traces. For the board with no plane, most of the heat is concentrated directly under and around the trace. As a plane gets closer to the trace layer, the surrounding area heats up and the "halo" spreads out dramatically wider. Thus, planes help cool the particular trace of interest, but they may spread the heat more efficiently to adjacent traces, with uncertain consequences.

6.3.7 Adjacent Trace: In most practical boards real estate is valuable. So there is almost always a trace nearby the trace we are concerned about. In this model we will add another 100 mil wide trace separated from the trace of interest by 8 mils. A diagram of this configuration is shown in Figure 6.20.

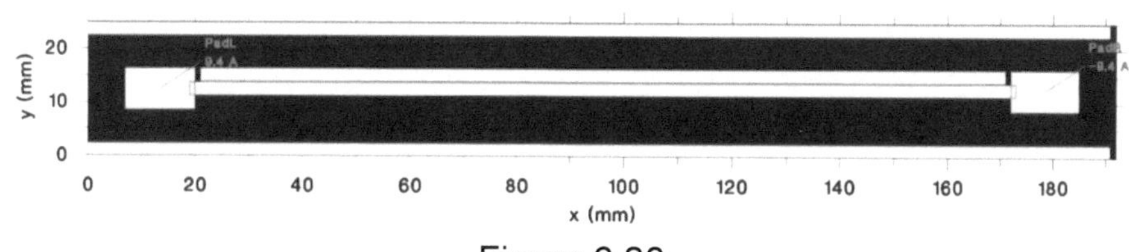

Figure 6.20
Adding an adjacent trace.

The result of this model, compared to the trace without a second adjacent trace, is shown in Figure 6.21. The temperature of the current carrying trace drops from 69.2 without a parallel trace, to around 62 to 65 when there is a parallel trace. BUT, the temperature of the parallel trace rises to around 55° C!

Thus, the temperature rise of one trace may have significant implications for traces surrounding it on the board.

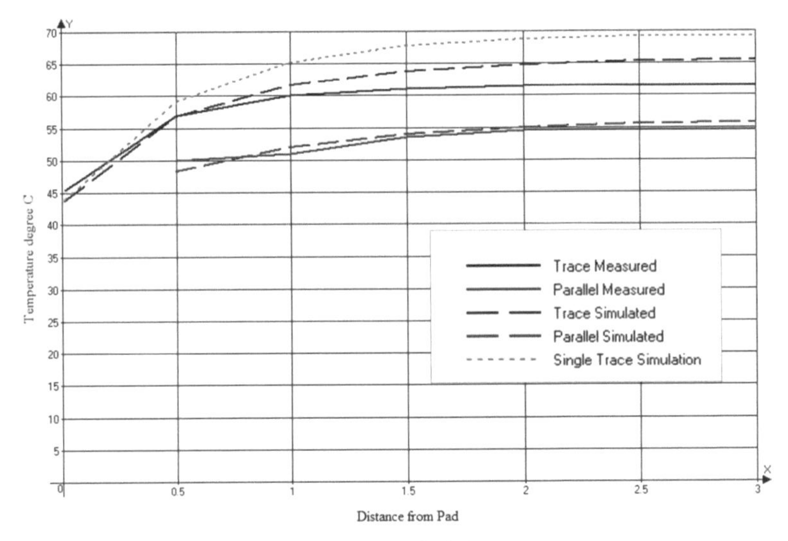

Figure 6.21
Effect of a parallel trace.

6.3.8 Adjacent Trace WITH Underlying Plane: When we added an underlying plane (Section 6.3.6), the temperature of the trace dropped to 44.9° C. When there was an adjacent trace (previous section) the adjacent trace heated to about 55° C. Now if we add an inner plane 8 mils under this structure, the temperature of the trace becomes 45.7 and the temperature of the parallel trace drops to 43.0 (See table 6.2). Thus, the biggest impact of adding a plane under the parallel structure is to help safeguard the second, parallel trace.

Condition	Driven Trace	Adjacent Trace
Single trace	69.2	
Single trace over plane	44.9	
Parallel trace	61.0	55.0
Parallel trace over plane	45.7	43.0

Table 6.2
Adding a parallel trace AND an underlying plane.

6.3.9 Dimensional Uncertainties: While designing the test board provided by Prototron Circuits, we set the nominal trace thickness for the traces at 1.9 mils. The 100 mil and 200 mil traces were located at one end of the board, while the 27 mil trace was located more toward the other end. I noticed a discrepancy in some of the data regarding the 27 mil trace (and those near it), so I went back and microsectioned the board. What we found was that the thickness of the 27 mil trace was actually more like 2.7 mils. This was a difference of around 40%.

Such differences are not uncommon on production-type boards. It is axiomatic that you cannot take the nominal trace dimensions as given when doing analyses. In fact, this is one of the larger problems in doing experimental analyses of boards --- knowing the *actual* dimensions you are dealing with. I tried pretty hard to make this point back in Section 3.4.

Table 6.3 is a comparison of the simulated results based on the two trace thicknesses.

Assumed thickness	1.9 Mil (Nominal)	2.7 Mil (Actual)
Temperature with 4.75 A	77.8	56.2

Table 6.3
Impact of dimensional uncertainty

6.3.10 Air Flow: Most of the simulations done in this chapter were done with HTC (Heat Transfer Coefficient) set to around 12 to 14 W/m^2K. Experimental calibration seems to suggest that this is a reasonable value. If we blow air across the trace, the effect is to increase HTC. But how to equate air flow to HTC when dealing with an isolated, small trace is extremely difficult to estimate [3]. Just for fun, I took the basic simulation model for the 27 mil trace and increased the HTC to 20 and to 28 W/m^2K. These values would simulate supplemental cooling, but how much and of what kind would be very uncertain. Therefore, the results are obviously to be taken with a large grain of salt. The results are shown in Table 6.4. It should be noted that HTC variations have greater impacts at higher temperatures.

HTC	Temperature
14	56.2
20	51.1
28	46.9

Table 6.4

Impact of changing Heat Transfer Coefficient.

6.3.11 Summary

Wow! Talk about confusing..... That was a lot of information. Table 6.5 is an attempt to summarize it all.

But when we get all through, what does it all mean? The bottom line is that this topic is very complicated and there simply are no easy answers. But there are a few generalizations we can make:

It is almost universally true that variations in results increase with temperature. That is, at low temperatures (associated with lower currents) the various parameters we have looked at here make little or no difference. As the temperature (current level) increases, changing these parameters has a correspondingly greater effect on the outcome.

The IPC curves generally represent a "worst case" scenario. Any other variation we introduce lowers the trace temperature, sometimes considerably so.

I have not addressed some of the more complex shapes, such as the fillets in thermal reliefs or the copper plating in drill holes. While we can speculate that the I^2R heating in these cases might be similar to other simulations, the cooling situations might be substantially more complicated.

Simulation Model	Temperature, Degree C	
	Basic Trace	Parallel or Adjacent
6"	69.2	
4"	68.0	
2"	60.9	
2" w/sink (1)	59.3	
1"	49.0	
1" w/sink (1)	46.3	
6" gradient (2)	44.8	
w/Opposite Plane	54.1	43.3 (4)
w/Underlying Plane	45.9	44.7 (4)
Parallel Trace	62.0	55.0 (5)
Parallel + Plane	45.7	43.0 (5)
With Air Flow (3)	46.9	
Notes:		
1. Heat sink under pads		
2. Measured adjacent to pad.		
3. *Very* approximate.		
4. Adjacent plane maximum temperature		
5. Adjacent trace maximum temperature		

Table 6.5
Summary of sensitivity simulations

The relationship between trace currents and temperatures is *very* complex. It is too complex to model with a single set of equations or curves. Since board real estate is very expensive, board designers usually want to use the smallest traces their fabricators will allow while still meeting the requirements. Optimizing board area when considering thermal effects does not seem possible without sophisticated thermal simulation software.

6.4 Sensitivities: Material Parameters:

In the previous section we looked at the sensitivity of the current/temperature relationship to various layout parameters. In this section we look at the current/temperature sensitivities to various material parameters and properties. To do so, we will develop a standard computer model to start with. That model is a

100 mil wide, 1.0 Oz. thick, 6 inch long trace on an FR4 dielectric (board) measuring 6.5 inches long by 2 inches wide. This dielectric is 63 mils thick, which is typical for modern PCBs. It will carry 7 Amps. The following parameters will be used in the base model. They have been found by the authors in previous studies to be typical and reliable.

> Resistivity: 1.68 uOhm-cm
> HTC: 11 (W/m^{2}K)
> FR4 with thermal conductivity coefficient: (W/mK)
>> In-plane .06
>> Through plane 0.4

Seven Amps of current is enough to raise the trace temperature to almost 60° degrees C, 40° above ambient. As noted above, sensitivities to layout and material parameters is a function of temperature increase, so this is a reasonable place to start from. We will then vary one parameter/property by itself and look on its effect on trace temperature. In effect, we are looking at the partial derivative of the current/temperature relationship with respect to that parameter.

6.4.1 Board thickness and planes: As the thickness of the dielectric increases, there is more board material for the heat to conduct through, so the temperature will presumably be lower with increased thickness. This is true up to a point, but there reaches a point where increased thickness does not buy you much. The typical PCB board is around 63 mils thick, and this is the thickness used for the basic model. But boards can be much thicker than that. Figure 6.22 illustrates the trace temperature as a function of board thickness. Note how the curve starts out fairly steep for thin boards, but then tends to stabilize as the board thickens.

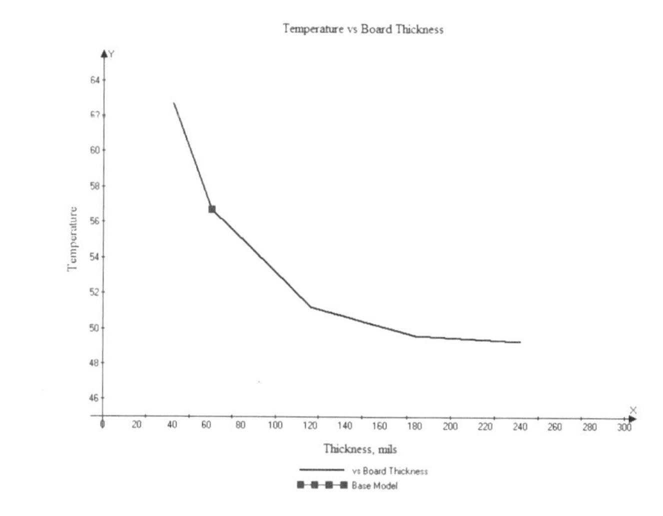

Figure 6.22
Trace temperature decreases as the board gets thicker, up to a point.

The dynamics associated with this effect can be seen in the thermographs shown in Figure 6.23 and 6.24. These show the temperature of the board material at and below the trace. Imagine a surface cutting through the middle of the trace (at the 80 mm point in the board) along the x-z plane. The view is towards the cross-section of the trace and shows the temperature gradient from the trace down toward the bottom of the board material. Figure 6.23 is for the 63 mil thick board, Figure 6.24 is for the 240 mil board. There is not enough material in the thinner board for the heat to spread out much, but that is quite different for the thicker board. That is why the trace on the thicker board is significantly cooler.

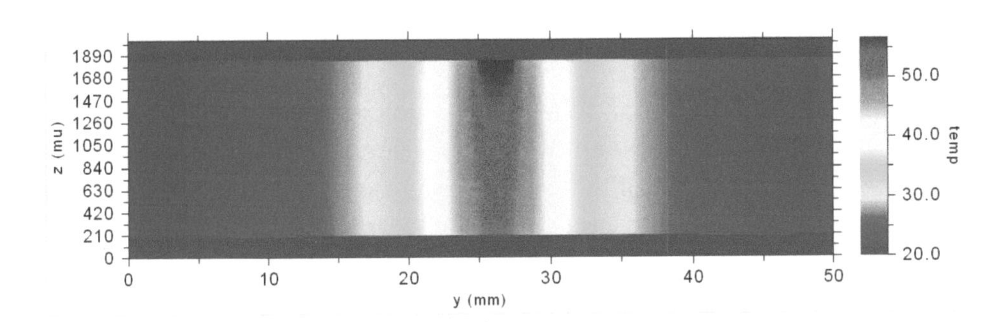

Figure 6.23
Vertical thermal profile of the 63 mil thick board.

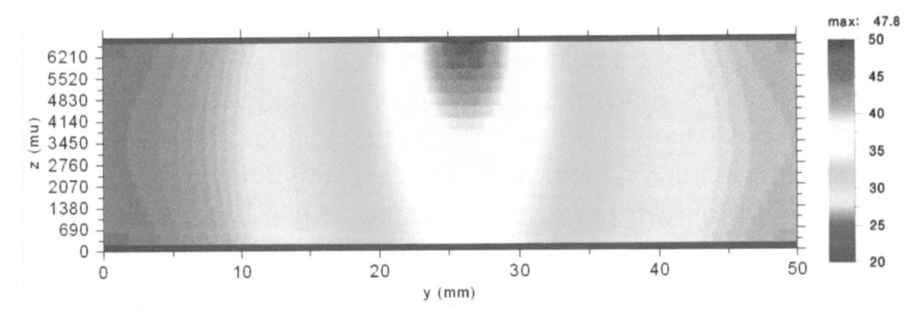

Figure 6.24
Vertical thermal profile of the 240 mil board.

Perhaps the largest external impact on trace temperatures can come from an adjacent plane. Many modern boards have internal planes, especially those with signal integrity issues. We did a simulation with a 63 mil thick board with an internal plane, 10 mils under the trace. This plane covered the entire surface area of that layer.

The impact of a plane is significant. It dropped the base temperature from 56.7 degrees C to 34.5 degrees C. But more than that, it spread the heat (and temperature) out much more widely than was the case without the plane. Figure 6.25 compares the thermograph of the trace layer with the underlying plane (b) to that of the trace without a plane (a). There is a dramatic difference.

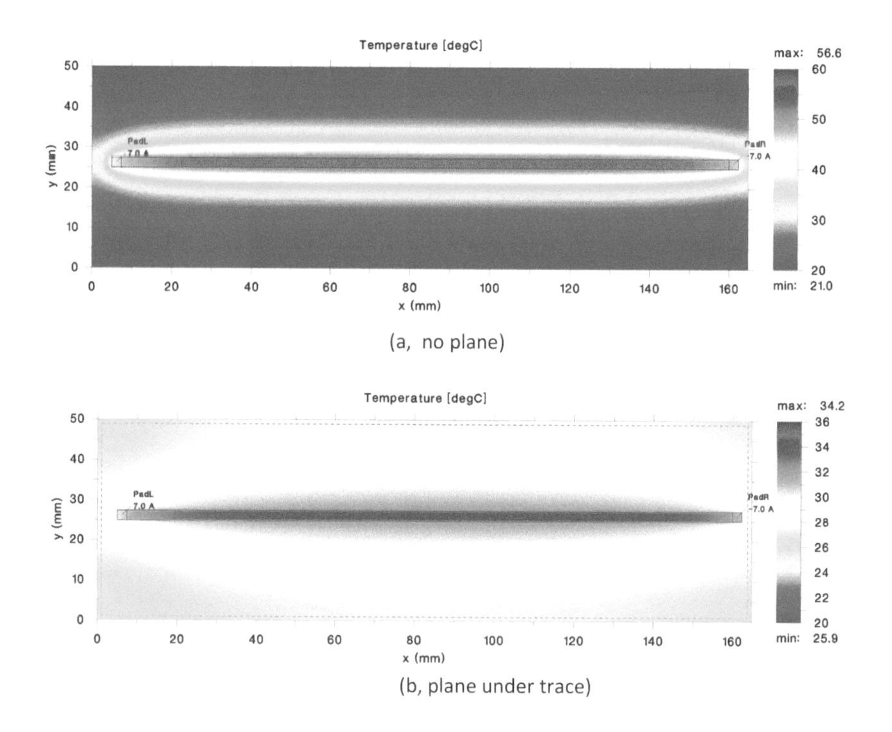

Figure 6.25
Impact of a plane on the thermal heat spreading.

The vertical distribution of the temperatures is also dramatically different. Figure 6.26 shows the vertical thermal profile for the case with the plane. Compare it with the case without a plane (Figure 6.23.)

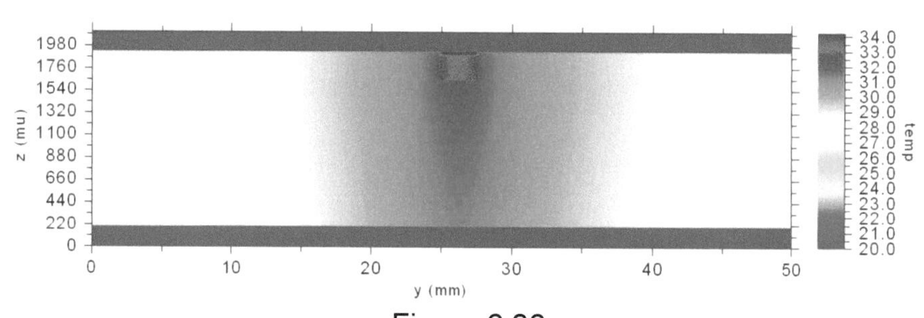

Figure 6.26
Vertical thermal profile with plane.

6.4.2 Effect of Resistivity: The resistivity of pure copper (at 20 degrees C) is 1.68 uOhm-cm. If the copper material on the board is plated (electrodeposited, or ED) copper, it will have a resistivity very close to this value [4]. Rolled copper, on the other hand, probably originates as a copper alloy, and has a resistivity anywhere between 1.72 and 1.78 uOhm-cm. Copper foil on PCBs can be either ED or rolled, but rolled is more common. The relationship between resistivity (p) of the material and the resistance (R) of the conductor is based on the geometry (area A and length L) of the conductor (see Equation 6.1).

[Eq. 6.1] $R = (\rho/A)*L$

Recall that the heating of the conductor is based on I^2R power dissipation in the conductor. Figure 6.27 shows the relationship between the trace temperature and the resistivity of the copper. Our standard model assumed "pure" copper. So the temperature goes up as resistivity varies from that. Note that the curve is linear, as would be expected.

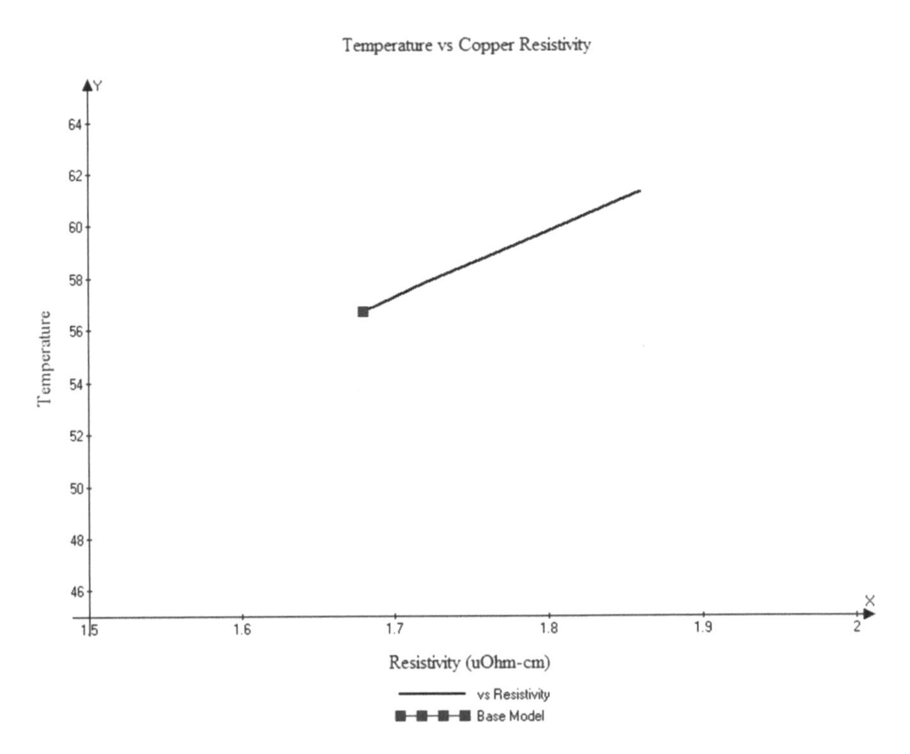

Figure 6.27
Trace temperature as a function of resistivity of the conductor.

6.4.3 Effect of Heat Transfer Coefficient (HTC): As mentioned above, HTC has components of convection and radiation. The authors have found in previous studies that for this temperature range, an HTC (W/m^2K) of 11 is pretty representative. An HTC of 6 would represent the situation of a trace in a vacuum (e.g. space). HTCs above 11 would represent some form of supplemental cooling, perhaps a heat sink or blowing air. We have looked at a range of HTC from 6 (vacuum) to 18, but there is no practical way today to say what 18 actually means! [5] Figure 6.28 is the relationship between trace temperature and HTC. The relationship shown in the figure is not surprising; as the convection cooling increases, the trace temperature goes down, and sometimes dramatically.

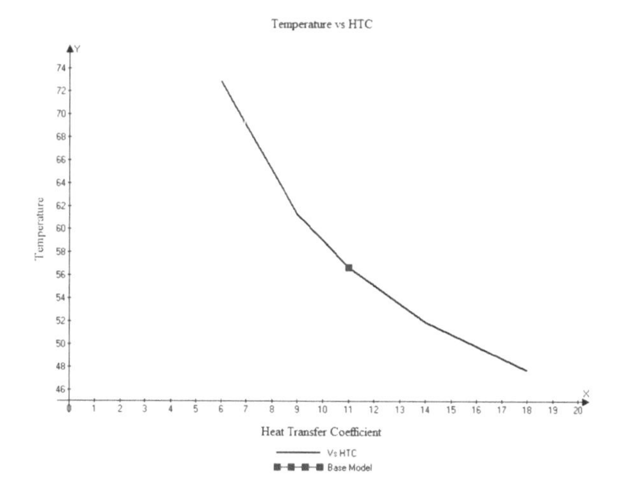

Figure 6.28
Trace temperature as a function of HTC

6.4.4 Effects of Thermal Conductivity Coefficient: Start with the idea that the board material has the greatest impact on trace cooling. Then with the idea that the thermal conductivity coefficient is the measure of how effective the board is at cooling the trace. Then recognize that this measure --- thermal conductivity coefficient --- is rarely provided in the data sheets of the material suppliers! And if it is provided, it frequently is provided as a single value, without specifying which coefficient it is. And if it is provided, it is often the value (through-plane) that is least important. This is an unfortunate state, indeed.

Figure 6.29 shows the relationship between the thermal conductivity coefficient and the trace temperature. The in-plane (x-y, or horizontal) coefficient is clearly the one that has more effect on trace temperature [6].

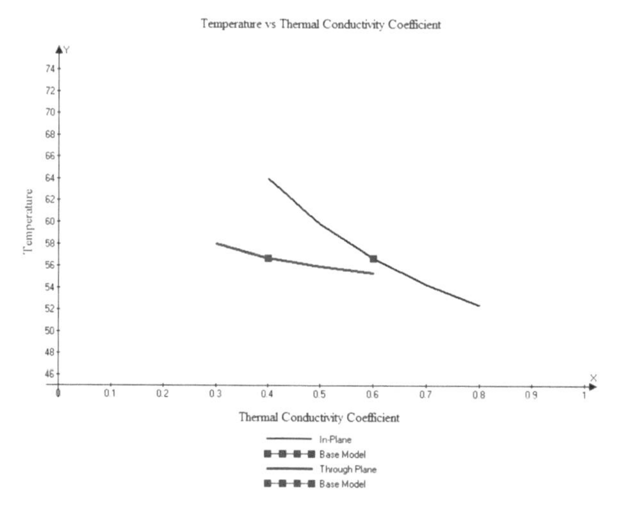

Figure 6.29
Trace temperature as a function of thermal conductivity coefficient.

6.4.5 Effect of Trace Thickness: Trace thickness is, of course directly related to cross-sectional area. So a change in thickness will have a significant effect on the current/temperature relationship. But some people might be surprised by how big an effect it can have.

Figure 6.30 shows the relationship. The red dot in the figure shows the results of the basic model, using a 1.0 Oz. thick trace. The curve of temperature vs, thickness is moderately steep in this range. For this model, the sensitivity is about 1.0 degree C for a change of 0.03 mil thickness. This is likely the reason traces do not heat uniformly (as discussed in Chapter 12.) Normal variations in trace thickness are will within this range.

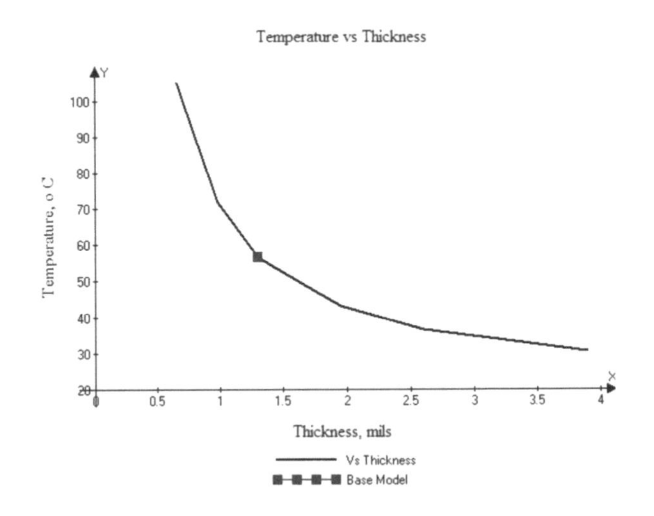

Figure 6.30
Trace temperature as a function of trace thickness

6.4.6 Summary:

The results of the simulations in 6.4.2 through 6.4.5 are summarized in Table 6.6. The values in the table, in effect, represent the partial derivatives of the change in temperature as a result in the change of one of the variables. While all the values are significant, perhaps the most sensitive parameter is thickness. We discuss the influence of thickness in more detail in Chapter 12.

Temperature Δ (Degrees)	For Each
1.0	2.0 % Δρ
3.0	0.1 mi. ΔTh
5.0	HTC between 11 and 14
5.6	Δ conduct. Coeff. 0.5-.07

Table 6.6
Sensitivity summary.

6.5 Voltage Drops Across Traces:

We can also use TRM to measure voltage drops across traces and vias. See more about this topic in Section 7.5

6.6 Role of Current Density:

There are some who believe that trace temperatures can be correlated to current density (A/in^2). Actually there is no direct relationship between current density and temperature. Here are two examples to illustrate the point.

Case 1: Assume two traces, A and B, are each carrying the same current. They each have the same cross-sectional area. Therefore the current densities are identical. But trace A is wide and thin and trace B is narrow and thicker. Trace A will be cooler because it has more surface area from which to dissipate heat.

Case 2: Assume two traces, A and B, are each carrying the same current and have identical cross-sectional areas and form factors. The current density will be identical in each trace. The resistivity of the conductor material making up trace A is higher than the resistivity of the conductor material making up trace B. Therefore the resistance of trace A will be higher than trace B. Therefore, the I^2R dissipation in trace A will be greater than in trace B and trace A will be hotter.

Conclusion: there is no necessary relationship between current density and temperature.

6.7 Conclusions:

Many board designers need to work hard to optimize their designs. That often means designing the narrowest traces they can or that their fabricators will allow. The reason is that board area is expensive. Narrower traces mean more traces per layer, which mean fewer layers. And layers are expensive.

So if a trace is carrying a significant current, the designer wants to design the narrowest trace possible, consistent with achieving his/her temperature constraints. All of the above suggests that can't be achieved with charts, equations, or formulas. Designers now need thermal simulation techniques to achieve their goals. And there is a precedent for this. In the 1990's many designers began to worry about controlled impedance traces [7]. At first we had equations that we could use, but in today's world, equations are not good enough. In the industry parlance, "field effect" solutions are required. Thermal simulation models, like TRM, employ the same quantitative techniques that "field effect" solutions employ.

That is not to say that there aren't some general conclusions we can derive. For example, looking at Figure 6.27, a 10.7% increase in resistivity leads to an 8.1% increase in trace temperature. It may pay to look at the copper characteristics of the materials used by the material supplier and the board fabricator. Similarly, if the thermal conductivity coefficient can vary by, say, 0.3 between material choices that may lead to a temperature reduction of as much as 10 degrees C [8].

6.7.1 Call to action: What is apparent from the above analysis is that there are two important areas where additional research is necessary. First, we do not know enough about how the HTC convection values change with temperature and with conditions. A trace represents a small hot spot on an otherwise moderately uniform plane. The industry needs to come up with ways to better estimate this effect.

Second, the industry needs to do a better job in providing thermal conductivity information about the materials that are supplied. But further, the authors hypothesize that (a) since the board material conducts heat in the in-plane direction more efficiently than it does in the through-plane direction, (b) therefore there should be a correlation between the ratio of resin to fiberglass and the trace temperature.

In addition, are there compounds that can be (economically) added to board materials (dielectrics) that can increase their thermal conductivity? For example, are there changes to the resin formulations that might accomplish this? Such approaches might improve the heat spreading capabilities of board materials almost to the point approximating the effects of underlying planes.

Notes:

1. Actually, the word "cube" is a misnomer. It is really a square column, the height of a layer. Nevertheless, I'll continue to use the word "cube" through this book.
2. If you would like to know more about heat transfer coefficients, start here http://en.wikipedia.org/wiki/Heat_transfer_coefficient
3. In fact, this would be a terrific avenue for further research.
4. The authors discuss how to measure the resistivity of copper traces, and the difference between ED and rolled copper, in Appendix 2.
5. Except it means that the trace is cooler!
6. For comparison, the thermal conductivity coefficient for copper is 385 (http://hyperphysics.phy-astr.gsu.edu/hbase/tables/thrcn.html)
7. Controlled impedance traces are those designed to look like transmission lines, so those lines can be terminated in their characteristic impedance, thereby reducing signal reflections back on the traces.
8. This would apply for trace temperatures in the range of 50 degrees or so. For higher trace temperatures, the difference would be even greater.

Part 3

SPECIAL TOPICS

7 VIA TEMPERATURES

7.1 Background Information

Through the years there has been very little written about the current carrying capacity of a via. This is, I believe, because there has been no practical way to measure or to predict the temperature of a via.

A via typically has a drill diameter and a copper conducting thickness formed during a plating process. This gives the via a conducting cross-sectional area. Consider the situation where a via connects a plane or trace on one layer to a plane or trace on another layer. If that via will carry a significant current (i.e. a current high enough to heat a trace to some degree), a relevant question is how big do we need to make the via's (conducting) cross-sectional area in order to safely handle that current? Until now there is been very little practical guidance on that question. But a thermal simulation model provides us with a tool to begin to explore that question.

Board designers typically approach this question using one of three strategies:

Simply don't allow vias that carry higher levels of current. Route all the conductors on the same trace layer.
Size the via to handle the current using the IPC 2152 guidelines.
Use a "standard" via known to be able to carry a certain amount of current and use multiple vias in parallel, as many as needed for the total current carried by the trace.

IPC 2152 explicitly endorses 2 and/or 3 on page 26:

> *The cross-sectional area of a via should have at least the same cross-sectional area as the conductor or be larger than the conductor coming into it. If the via has less cross-sectional area than the conductor, then multiple vias can be used to maintain the same cross-sectional area as the conductor.*

Anecdotally this must be good advice because most designers have never had a via fail for purely thermal reasons (unless there was a fabrication or alignment issue.) What we have been doing must be right, because it works so well! In fact, this guidance has been *extremely* conservative.

But there is an implicit assumption in this guidance that sometimes poses a problem. That is, it is assumed we know the plating thickness of the via. We typically assume (*hope* might be a better word) that the via wall thickness is the same as the plating thickness, which is typically 0.5 or 1.0 Oz. copper (0.65 to 1.3 mils). But the plating is often not uniform on the walls of the via. This is a separate problem that I am not prepared to address in this book.

7.2 Thermal Simulation

In Chapter 6 I described the basics behind a thermal simulation model. Think of the board being modeled as consisting of a large number of small cubes. Each cube is small enough so that the parameters don't change much across any of its surfaces. So the computer model solves for the boundary conditions of each cube, and then starts assembling a complete model.

The sides of the cubes must be smaller than the smallest dimension (in the x,y planes) on the board being modeled. For a via simulation, this is the thickness of the conducting surface on the wall of the via. This is a much smaller dimension than most models have, and indeed smaller than any models developed so far in this book. The problem is that such small cubes means that there are LOTS of cubes making up the model [1]. That, in turn, means the matrices are very large, and that means the CPU load on the computer is very large (meaning long processing times.)

The practical response to this issue is to model a board that is smaller than what a conventional board might look like. This will (probably) result in higher than actual temperatures, but not too much higher. More importantly, any *relationships* between and among various modeled traces will be preserved.

7.2.1 Simulation Strategy: So we are going to proceed with the following set of assumptions and strategies:

1. We will assume a standard via with a 10 mil (0.26 mm) drill diameter and a 0.030 mm uniform plating thickness. This is slightly less than a 1.0 Oz equivalent. The effective cross-sectional area of this via is $\pi(r^2 - (r\text{-}th)^2)$ where r is the via radius and th is the plating thickness. Choosing these mm equivalents (which are slightly rounded) results in a via cross-sectional area of .0217 mm^2. This is approximately equivalent to a 1 Oz. 26 mil (.66 mm) wide trace.

2. The pad area of a via doesn't matter because we are going to be dealing with traces that are at least 26 mils (0.66 mm) wide. (If we were dealing with smaller traces, the via cross-sectional area would exceed that of the trace and should be OK from a heating standpoint "by definition.")
3. We will be dealing primarily with a standard board thickness of 63 mils (1.6 mm).
4. We will vary the trace width and current and compare the temperature of the trace on the board with the via temperature. In this way, we will see if the via gets hotter than the trace, stays the same temperature as the trace, or runs cooler than the trace.
5. We assume that at low currents via temperature is not a problem. Since we are primarily concerned with high-temperature effects, we will concentrate on what happens at higher temperatures and currents.

7.2.2 Board Model: Our thermal model is a pair of 1.0 Oz., 60 mm long traces on either side of a board connected by a 0.26 mm diameter, 0.03 mm plated via. The board is a few mm wider than the trace. The dielectric layer between the trace layers is 1.6 mm FR4. A thermocouple (TC in Figure 1) is placed near the midpoint of the top trace. This will measure the temperature at that point during the simulations. The purpose for this will be described later. A diagram of this model is shown in Figure 7.1.

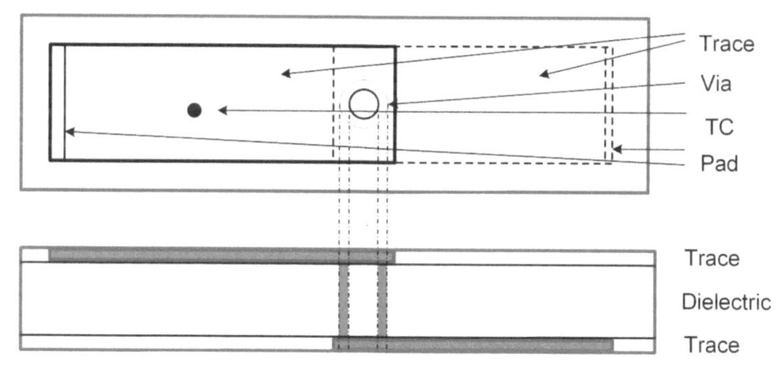

Figure 7.1
Thermal model of trace with via. *(Not to scale)*

7.2.3 First Simulation: My first series of simulations will be run with this model in five widths, 0.40 mm, 0.66 mm, 1.2 mm, 1.8 mm, and 2.5 mm. (This approximately corresponds to widths of 15 mil, 26 mil, 47 mil, 71 mil and 100 mil, respectively.) The results of the first simulation, 0.66 mm, are shown in Table 7.1. This is the simulation where the via and trace conducting areas are approximately equal. The values shown are the temperatures (NOT the temperature *change*) at the thermocouple (TC, the maximum temperature point of the upper trace), the top of the via, the midpoint of the via, and the "Bare trace."

Temperatures in the Simulation at Various Points and Currents				
Current (A)	TC	Via top	Via midpoint	Bare Trace
3	86.9	81.4	80.9	86.2
4	166.7	154.1	152.9	165.0

Table 7.1
1.0 Oz. , 0.66 mm trace with 1.0 Oz. via.

One of the first questions to ask about the simulation temperatures is "compared to what?" I also modeled each configuration as a single trace ("bare trace") on a single layer with no via, under the same conditions. This way I can compare the via temperatures with what the trace temperature would be without the via. The thermocouple (TC) was intended to provide this information. The "Bare Trace" temperature is a check on that assumption. In every simulation the "Bare Trace" temperature was nearly identical to the TC temperature, normally differing by significantly less than one degree C. Only above 100 degrees C did the difference approach or slightly exceed 1.0 degree C.

In this first simulation, the via cross-sectional area and the trace cross-sectional area are approximately the same. Under the assumption that the cross-sectional area of the via determines the via temperature, the via midpoint temperature should equal the same as the TC (which equals the temperature of a "bare trace" without a via.) We can see from Table 7.1 that *this is not true*. At 3 Amps the via runs approximately 6 degrees C *cooler* than the trace and at 4 Amps it runs approximately 12 to 13 degrees C *cooler*. The thermal profile of the top trace layer at 3 Amps is shown in Figure 7.2. It is apparent that the trace begins to *cool* as approaches the via in the center.

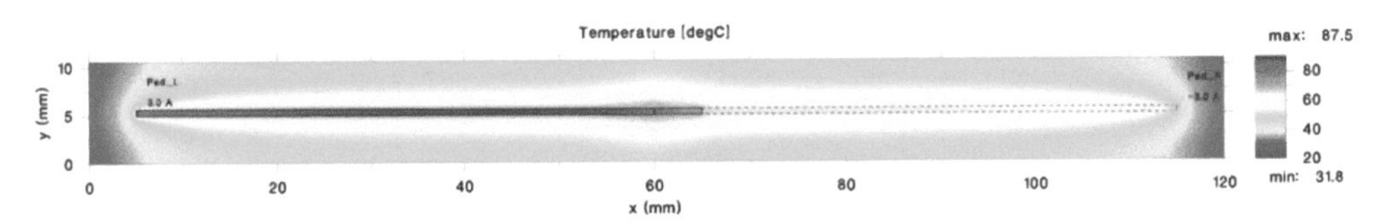

Figure 7.2
Thermal profile of the top trace layer of the first simulation at 3 Amps.

The results of the first simulation are shown graphically in Figure 7.3. In these graphs the via midpoint temperatures are shown in red and the TC temperatures are shown in black. All traces are assumed to "start" at 20 degrees C at 0.0 Amps. Note that the via temperature is lower than the trace midpoint temperature.

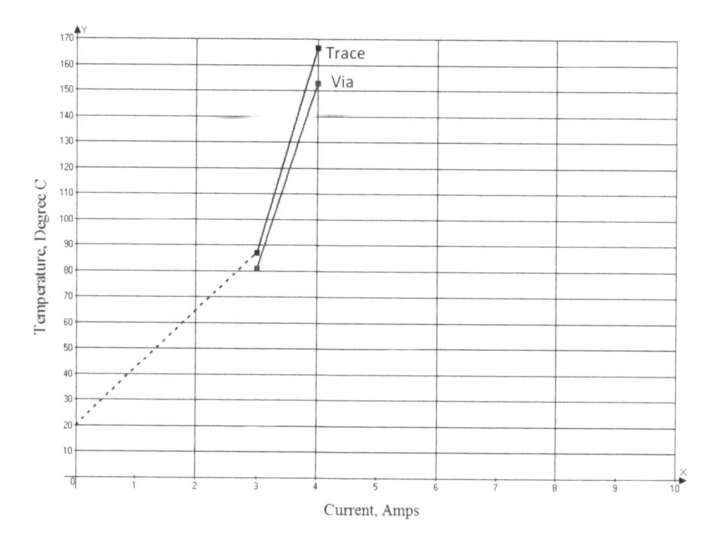

Figure 7.3
Via and trace temperature vs. current for
a 0.66 mm wide 1.0 Oz. trace.

7.2.4 Additional Simulations: I ran additional simulations at four other widths. The results for all five simulations are shown in Table 7.2, and graphed in Figure 7.4. The graph sets from left to right are for 0.4 mm, 0.66 mm, 1.2 mm, 1.8 mm, and 2.5 mm traces respectively (approximately 15 mil, 26 mil, 47 mil, 71 mil, and 100 mil.)

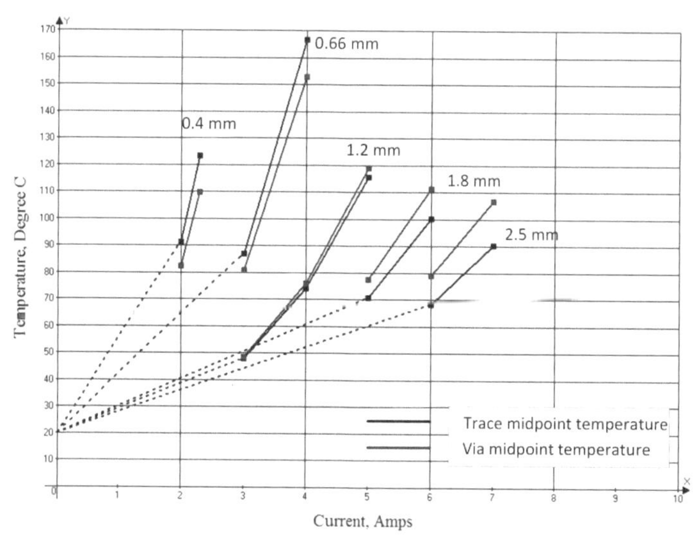

Figure 7.4
Via (red) and trace (black) temperature vs. current
for selected widths.

Temperatures in the Simulation at Various Points and Currents				
Width (mm)	Current (A)	TC	Via top	Via Midpoint
0.4	2	91.4	83.2	82.4
0.4	2.3	123.2	111.1	109.9
0.66	3	86.9	81.4	80.9
0.66	4	166.7	154.1	152.9
1.2	3	47.9	48.6	48.7
1.2	4	74.1	75.7	75.8
1.2	5	115.5	118.75	118.75
1.8	5	70.8	76.8	76.8
1.8	6	100.2	110.2	110.2
2.5	6	68.1	77.6	77.6
2.5	7	90.3	104.9	104.9

Table 7.2

What is apparent from the data is the following:

1. If the conducting area of the trace is smaller than that of the via, the trace is hotter than the midpoint of the via.
2. If the conducting area of the trace is larger than that of the via, the trace is cooler than the midpoint of the via.
3. Therefore, there must be a trace conducting area where the trace temperature and the via midpoint temperatures are equal. The simulations suggest that this occurs when the trace conducting area is approximately 1.5 to 2.0 times the via conducting area.

Looking at the first simulation (0.66 mm), the reason that the via runs cooler than the trace (for the same conducting areas) has not been obvious up until now. One of the surprises in IPC 2152 is that internal traces run somewhat cooler than do external traces of the same dimension. That is because, it turns out, the heat spreading through the dielectric supports more cooling than does convection and radiation into the air. The via looks very similar to an internal trace surrounded by a dielectric. That is why it runs cooler than an equivalent trace on the top layer. Furthermore, the copper is an extremely good conductor of heat. The via and the trace have the same cross-sectional area, and therefore the same resistance per-unit-length. Therefore, we can expect them both to heat approximately equally through the I^2R heating mechanism. Since the via runs cooler than the trace, what is happening is that *the via is helping to cool the trace.*

As the trace width increases, and as the current increases, the via does start running a little hotter than the trace, but only by a few degrees. As the trace width gets wider, and therefore can handle the thermal heating better than the via can, the trace starts cooling the via (because of the high thermal conductivity between the via and the trace.) For example, look at the comparison in Table 7.3.

Temperatures in the Simulation at Various Points and Currents				
Width (mm)	Current (A)	TC	Via top	Via Midpoint
.04	2	91.4	83.2	82.4
0.66	3	86.9	81.4	80.9
1.2	4	74.1	75.7	75.8
1.8	5	70.8	76.8	76.8
2.5	6	68.1	77.6	77.6

Table 7.3

The via cross-sectional area is about the same as that for a 0.66 mm trace. Under the old guidelines, we would expect the via temperature to increase with current. But the via midpoint temperature stays approximately the same as we increase the current and the trace width at the same time. At 6.0 Amps the via is running *cooler* with a 2.5 mm trace than it does at 2 Amps with a 0.4 mm trace!

The comparison is even more dramatic if we simulate a 2 Oz. trace. Just for fun, I look at a 2.5 mm, 2 Oz. trace carrying 7 and 10 Amps with only a single via. The results are shown in Table 7.4. Note that with 10 Amps the via midpoint temperature is about 118 degrees C, almost 25 degrees hotter than the trace TC. But consider this: a single 0.66 mm trace (with the same cross-sectional area as this via) carrying 10 Amps would reach the melting point of copper (1083 degrees C) in approximately 7.5 seconds [2]! The via only reaches 118 degrees C because of the thermal coupling to the 2 Oz. trace.

Temperatures for 2.5 mm, 2 Oz. trace, single via			
Current (A)	TC	Via top	Via Midpoint
7	51.3	59.0	60.8
10	93.5	113.2	117.9

Table 7.4

All the above simulations used enough current to raise the temperatures (not the change in temperatures) to a pretty high level. This was to exaggerate the results. Most designers would not design to these levels. Typically designers target for only a 20 to 40 degrees C temperature rise maximum. But what if we use our standard via (designed for a one ounce, .66 mm trace) with a 2.0 Oz. trace with only enough current to raise the temperature by 20 or 30 degrees C.?

I ran one final simulation of a 2.0 Oz., 2.5 mm wide trace raised to a relatively modest temperature. The results are shown in Table 7.5. At 7.0 Amps, the via midpoint temperature is almost 41 degrees C above ambient (of 20 degrees C, see Table 7.4), while the trace is about 31 degrees above ambient. At 5.0 Amps (Table 7.5), the trace is only 17.1 degrees above ambient and the via just slightly higher at 21.1 degrees above ambient. This is in spite of the fact that the same via with only 4.0 Amps is 146 degrees above ambient (see Table 7.2) in a 0.66 mm trace with the same conducting area as the via!

Temperatures for 2.5 mm, 2 Oz. trace, single via			
Current (A)	TC	Via top	Via Midpoint
5	37.1	40.3	41.1

Table 7.5

7.2.5 Two Vias: Even though a single via can carry significantly more current than expected, as a result of the thermal coupling to the trace, the capacity is of course greater if we use two vias. Table 7.6 provides the results of simulations of two different 2.5 mm (100 mil) traces each with two vias. One trace is a 1.0 Oz trace and the other is a 2.0 Oz trace. The temperatures at the two vias on each trace are virtually identical, confirming the fact that the current is splitting equally between the two vias. The graphical results are shown in Figure 7.5. The blue lines are the via midpoint temperatures and the green lines are the trace TC temperatures.

Note that even at 10 Amps through a 2 Oz. trace, the small vias are less than 10 degrees warmer than the trace.

Temperatures for 2.5 mm, 2 Oz. trace at Various Points and Currents, Two Vias					
Weight (Oz)	Width (mm)	Current (A)	TC	Via top	Via Midpoint
1.0	2.5	6	67.1	69.5	69.5
1.0	2.5	7	88.6	92.3	92.4
2.0	2.5	6	41.3	43.3	43.5
2.0	2.5	7	49.9	52.7	53
2.0	2.5	10	89.6	96.5	97.3

Table 7.6

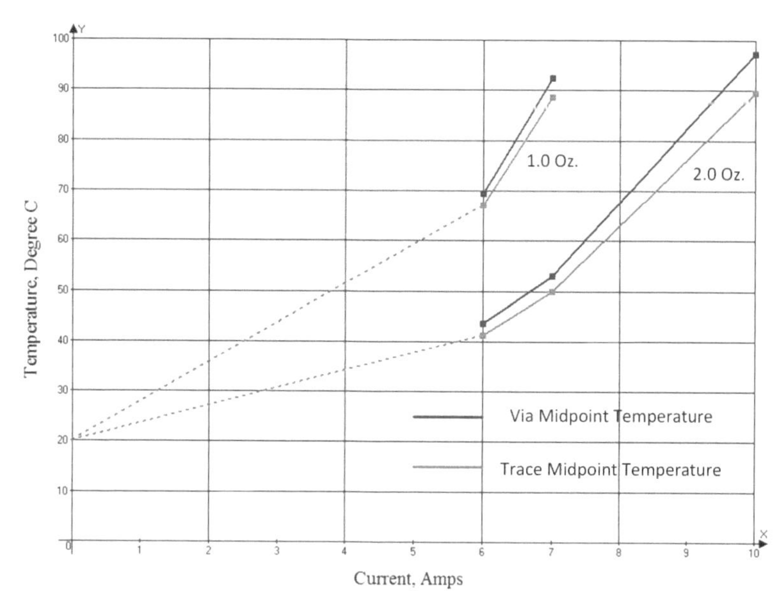

Figure 7.5
Trace and via temperatures for a 2.5 mm (100 mil) wide trace with two vias.

7.2.6 Thermal Vias: There are designers who use thermal vias to connect a heat sink to a plane or to connect two planes together for thermal coupling reasons. This is a perfectly legitimate purpose for a via. Note that in this case the via is not carrying any significant current at all (at least it shouldn't be). It is simply a path for heat to travel to another surface where it can more readily be dissipated. The foregoing shows why thermal vias can be so effective. The thermal conductivity is extremely good and effective through a via. In fact, what the previous sections show is that thermal vias do not have to be very big nor very numerous to perform their function.

7.2.7 Conclusion: The inescapable conclusion from the above is that the conventional wisdom has been wrong! The current does not determine the temperature of the via, the associated traces do. As long as the traces are sized to properly handle the current, even a single, ordinary via is adequate to transition between trace layers. And the reason is because of the thermal conductivity of the copper between the via and the trace.

If there was ever a set of conclusions that cried out ***"SHOW ME!"*** these are they!

7.3 Experimental Verification

To test the results of the previous section, the first thing we need is a test board. Prototron Circuits (Redmond, WA. And Tucson, AZ.) [3] was very generous in providing a test board for this purpose. The relevant portion of the test board is shown in Figure 7.6.The board is approximately 60 mils thick FR4.The board contains 0.5 Oz. copper nominally plated with 1.0 Oz. additional copper. Two traces were compared, one nominally 27 mils wide, the other 200 mils wide. Each trace is 6.0" in length, one-half on the top layer and one-half on the bottom layer. Each trace has a single 10 mil diameter plated via connecting top to bottom. It is important to note that the via structure is identical in each trace. The board was supported 2.5 inches above a plywood surface in still air by four screws at the corners. The board was microsectioned after testing to measure the actual dimensions for the simulation [4].

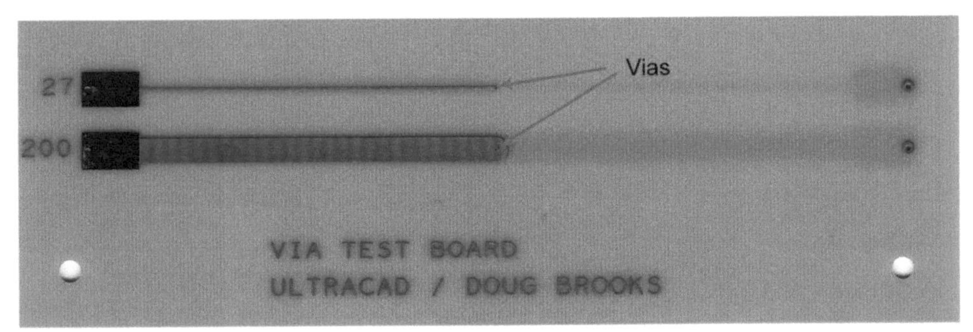

Figure 7.6
Relevant portion of via test board

The 10 mil diameter via plated to 1.0 Oz. has roughly the same conducting cross-sectional area as the 27 mil trace. That is why the 27 mil wide trace was chosen for this study.

7.3.1 Simulation: Before I tested the actual board, I felt it was important to replicate the simulation results on the actual board used for testing. For the reasons described above (Section 7.2), the simulation model has to be smaller than the actual board, leading to the possibility of slightly higher simulation results than actually measured. It turned out there was one other measurement that was necessary, determining the *actual* thermal conductivity of the board material. C-Therm Technologies [5] (Fredericton, New Brunswick) graciously measured the thermal conductivity of the board material to facilitate the simulation. They measured the thermal conductivities as (W/mK):

In-plane	0.679
Through-plane	0.512

These thermal conductivity values are significantly higher than those modeled in Section 7.4, which will result in lower overall temperatures. Also, in Section 7.2, mor modeled trace and the via wall had the same thickness (1.0 Oz.). The actual test board has plated copper over 0.5 Oz. foil, so the nominal trace thickness will be 0.5 Oz. thicker than the via wall. This will also lead to a lower overall temperature.

7.3.2 Simulation Results: Table 7.7 shows the result of the simulation. They are consistent with the type of results presented in Section 7.2, above. For a small trace, approximately the same size as the via, the via is cooler than the trace because the via cools more effectively into the dielectric layer of the board than the trace does. For the larger trace carrying more current, the heat generated in the via is conducted away from the via onto the trace, resulting in the via being a little hotter (but not a LOT hotter) than the trace.

Result of Via Temperature Simulation				
Trace width (mils)	Current (A)	Trace Temp. oC	Via Temp. oC	Via T/Trace T
27	4.75	72.8	70.1	96.3
27	6.65	114.2	108.2	94.7
200	4.75	30.8	31.8	103.2
200	8.55	44.8	48.1	107.4

Table 7.7

Figures 7.7 and 7.8 are the simulated thermal profiles of the 200 mil trace carrying 8.55 Amps and the 27 mil trace carrying 6.65 Amps, respectively. Note that the thermal profile cools a little near the via for the 27 mil trace. The via is the hottest point on the 200 mil trace, but only 12% hotter than the trace itself.

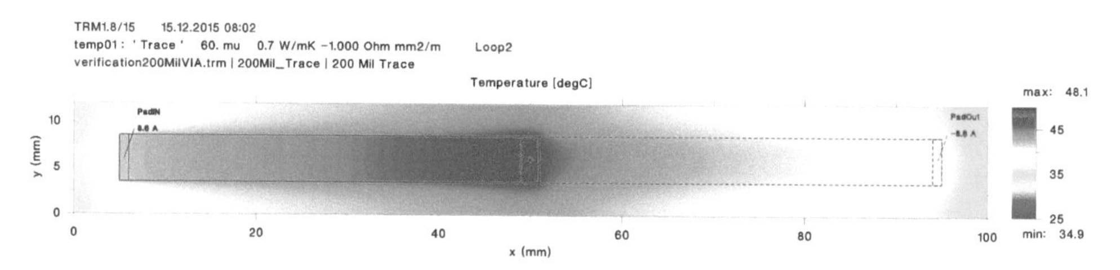

Figure 7.7
Thermal profile of top layer of 200 mil trace carrying 8.55 A.

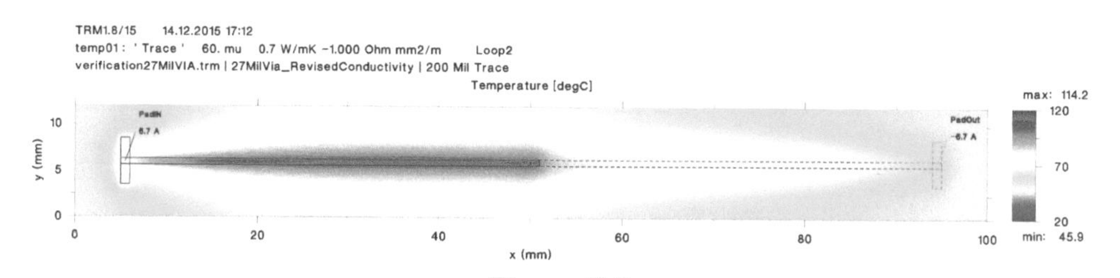

Figure 7.8
Thermal profile of top layer of 27 mil trace carrying 6.65 Amps

7.4 Experimental Results

The test procedure is pretty straightforward. A constant current is applied to the trace. The current level is checked and confirmed by a second digital meter. The temperature is measured with a thermocouple probe whose tip diameter is spec'd at 30 Ga. (approximately 10 mil diameter). The probe response time is approximately one second. The probe was calibrated to an ice cube and to boiling water at an elevation of 360 feet and found to be perfectly calibrated. Various tests were run to verify that the probe does not affect the temperature of a trace being measured.

For each test, a constant current is applied to the trace until the temperature stabilizes (approximately six minutes.) Then the temperature is measured and recorded. After the measurements were taken, the board was returned to Prototron for microsectioning to confirm all dimensions.

7.4.1 Measured Results: The measured test results are provided in Table 7.8

There are two especially important observations that should be made here:

1. First, a 6.6 Amp current through the 27 mil wide trace results in a via temperature of 109° C while a *higher* current of 8.6 Amps in a larger (200 mil wide) board results in a ***much*** lower via temperature of only 44.5° C. This confirms that it is the trace that is controlling the via temperature.
2. The measured data (Table 7.8) are very close to the simulated data (Table 7.7.) This gives us confidence that the simulation approach is a viable approach for predicting temperatures in complex situations.

Measured Results of Via Tests				
Trace width (mils)	Current (A)	Trace Temp. oC	Via Temp. oC	Via T/Trace T
27	4.75	66	64.5	97.7
27	6.65	114	109	95.6
200	4.75	30.5	31.5	103.3
200	8.55	40.5	44.5	109.9

Table 7.8

Figures 7.9a and 7.9b show thermographs [6] of the 27 mil and 200 mil traces carrying 6.55 Amps and 8.55 Amps, respectively. Note the similarity to the simulated thermal profiles. In particular, Figure 7.9a shows the hottest part of the trace well to the left of the via, consistent with the above results. Figure 7.9b shows the hot spot at the via, but only a few degrees hotter than the rest of the trace, ***AND*** much cooler than the via in Figure 7.9a (even though it is carrying more current.

7.4.2 Conclusion: The results of this experimental evaluation are consistent with, and would seem to confirm, the results predicted in Section 7.2.7. It is not the current that determines the temperature of a via, it is the temperature of the associated traces. If the traces are sized correctly for the current level, *MUCH* smaller, and fewer, vias are required to transition between trace layers than has been previously believed. This means designers can have much greater flexibility in freeing up routing channels underneath the traces on their boards.

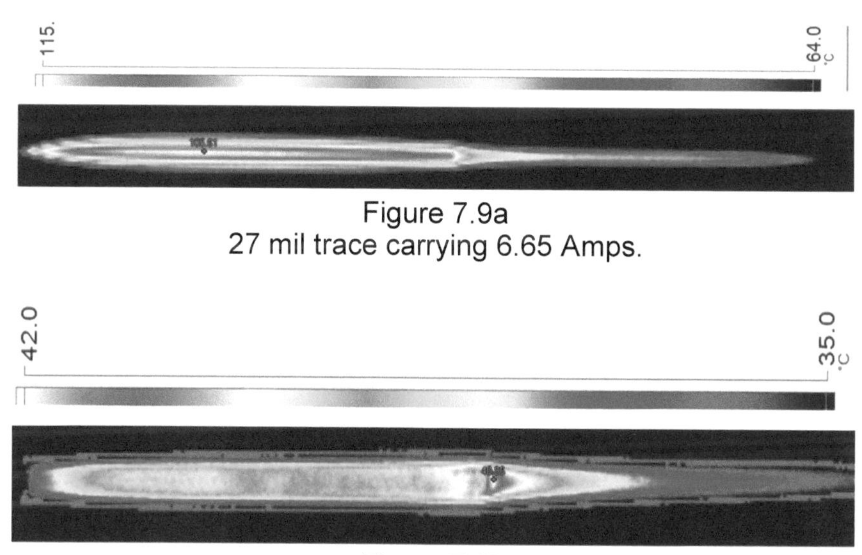

Figure 7.9a
27 mil trace carrying 6.65 Amps.

Figure 7.9b
200 mil trace carrying 8.55 Amps.

7.5 Voltage Drop Across Trace and Via:

Designers and engineers are sometimes concerned about the voltage drop across a trace or a via. TRM allows us to analyze this fairly easily. We will use the 26 mil (.66 mm) trace we modeled with 3 Amps of current in Section 7.2.3.

We modeled this trace as well as a trace with identical dimension but no via (the "bare trace" shown in Table 7.1), using the latter as a control. TRM allows us to look at the voltage gradient at any point along the trace (just as it allows us to look at the thermal gradients, as we have seen, and at the current gradients, as we will see in the next chapter.) The voltage drops we derived using TRM were:

.66 mm wide trace without via:	0.34 Volts
.66 mm wide trace with via:	0.32 Volts
Volts across via:	0.00 Volts

Interestingly, TRM measured ***no*** voltage across the via and ***less*** voltage across the trace ***with*** a via than the trace ***without*** the via.

The reason the voltage across the trace with the via is smaller is because the trace with the via is cooler than the trace without the via (see Table 7.1). That is because the via is helping to cool the trace. The reason there is "no" (actually a very small) voltage across the via is because the via is so short.

We can check some of these figures using Equation 3.1 and Equation 3.5 in Chapter 3, reproduced here as Equations 7.1 and 7.2.

[Eq. 7.1] $R = (\rho/A)*L$
[Eq. 7.2] $R(T) = R(T_o)*(1+\alpha_o*(T - T_o))$

First, let our parameters be:

ρ = 1.7 uOhm-cm
α = 0.004
T_0 = 20 degrees C
T = 86 degrees C
L = 115 mm (4.5 inches)
W = .66 mm (26 mils)
Th = 1.0 Oz

These parameters lead to an initial resistance of 0.0897 Ohms at 20 degrees C and a final resistance of 0.11339 Ohms at 86 degrees C. Then, using Ohms Law, we see that the calculated voltage drop would be 0.11339 * 3 = 0.34 volts, just as TRM predicts.

Then, the voltage drop across the trace, 0.34 Volts, occurs across 115 mm. But our via length is only 1.6 mm (63 mils), the thickness of the board. So we can calculate the voltage drop across the via as:

$V_{via} = 0.34 * 1.6 / 115 = 0.0047$ Volts

That is, the voltage drop across the via is negligible compared to the voltage drop across the entire trace.

In Section 7.2.4 we also modeled a 100 mil (2.5 mm) wide trace with 7 Amps, reaching a trace temperature of 90 Degrees C, slightly higher than the 26 mil wide trace with 3 Amps. The resistance and voltage drop for that trace (without a via) would be (using Equations 7.1 and 7.2):

R at 20 degrees C	0.0237 Ohms
R at 90 degrees C	0.0303
V = .0303 * 7 =	0.212 Volts

Assuming the via remains the same resistance (in this illustration it would be slightly higher) we can estimate the voltage drop across the via by using proportions:

Via voltage (at 7 Amps) = .0047 * 7 / 3 = 0.011 Volts

This is still substantially below other voltages along the trace.

Summary: The voltage drop across a trace with a typical via is lower than we might expect, based on the maximum temperature of the trace, because the via helps cool the trace. And since the length of a via is generally much less than the length of the trace, the additional resistance of the via contributes a negligible increase in voltage drop. In general, from a practical standpoint, we can consider a voltage drop across a via to be insignificant compared to the voltage drop across the traces.

Notes:

1. This is related to the "Thermal Pixel" size referenced in Section 6.2.
2. See Chapter 10. The calculation was made with UltraCAD's UCADPCB4 Trace Calculator, available at www.ultracad.com.
3. Contact: Dave Ryder, Prototron Circuits, Inc., 15225 NE 95th St., Redmond, WA 98052, www.prototron.com .
4. The trace widths were essentially as designed. The thicknesses were (top layer) 2.1 mil and (bottom layer) 2.9 mil.
5. See their measuring process described in Appendix 1. Contact: Adam Harris, C-Therm Technologies, 921 College Hill Rd, Fredericton, NB, Canada, E3B 6Z9, www.ctherm.com .
6. For more on thermographs see Chapter 12.

8 VIA CURRENT DENSITIES

While Johannes and I were investigating the relationship between current and via temperature, we noticed a very interesting pattern in the current density at various points within a via. We ran a separate series of simulations designed to look more closely at current density

8.1: Single Via

Our first simulation was of a simple trace on the top layer of a board connecting to a trace on the bottom layer by means of a through-hole via. The trace was .33 mm (13 mil) wide and 34 um (1 Oz.) thick. The board dielectric was FR4, 1.61 mm (63 mil) thick. The board itself was a few mm wider than the board. For simulation purposes, the dielectric was modeled as a set of seven individual layers, each 230 um thick. Using multiple layers in the simulation allows better precision in the model, as well as allowing us to look at the current densities at various points along the via wall. The via was .26 mm (10 mil) wide with a wall thickness of .03 mm (slightly less than a 1.0 Oz. equivalent.) This results in a via conducting area of .021677 mm^2. See Figure 8.1.

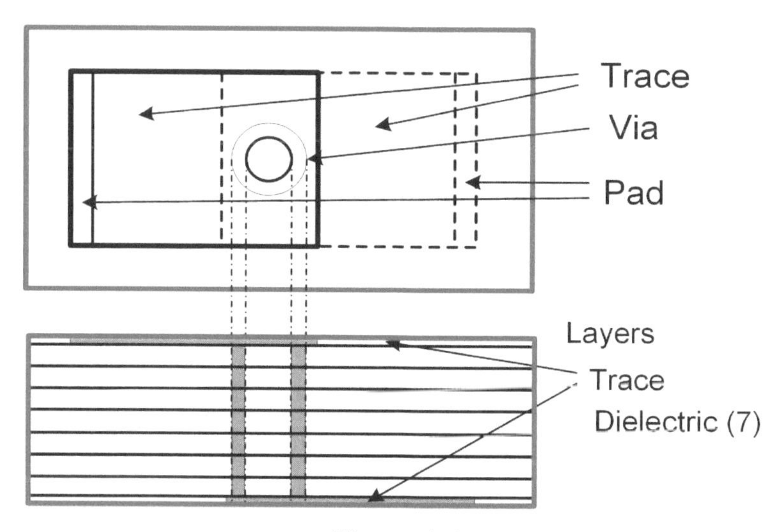

Figure 8.1
Simulation model of a single via.
(Not to scale)

The current density patterns of the first two layers (trace layer and Core 1) are shown in Figure 8.2 for an applied current of 4 Amps. Appendix 6.1 offers a detailed explanation of how to interpret these figures, and Appendix 6.2 provides the full set of patterns through Core 4. I encourage the reader to read Appendix 6.1 before proceeding further in this chapter. I will look at the current density pattern for the Core 1 layer first, then come back to the top (trace) layer.

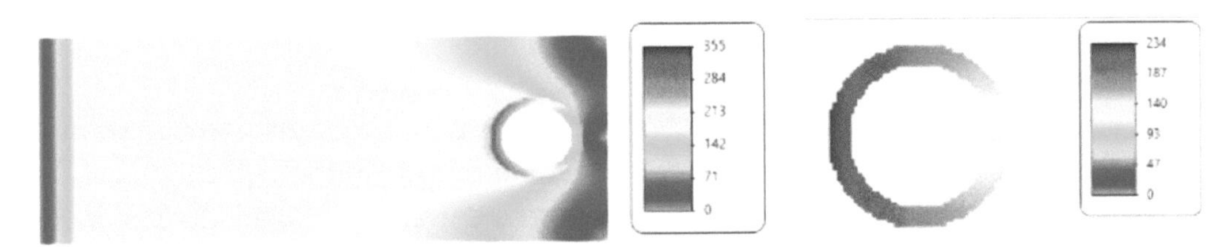

Figure 8.2
Current densities (Amp/mm^2) on the top, trace layer (left)
and the first core layer (right).

Figure 8.3 shows the current density pattern of the Core 1 layer in more detail. The current density (in A/mm^2) varies from 223.7 on the near (leading) side to 140.4 on the far side of the via wall. If we total these densities we get 723.8, which, if divided by the four measurements, results in an average density of 181.0 A/mm^2. If we then multiply this average density by the area of the via wall (.021677 mm^2) we get 3.93 Amps.

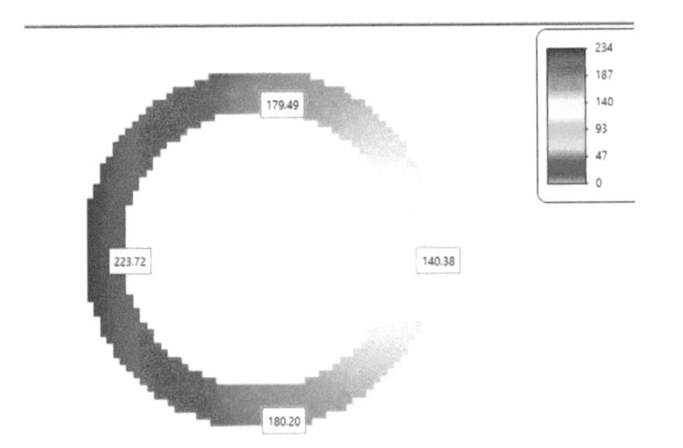

	density	Fraction of total
1	223.7	0.308
2	179.5	0.248
3	140.4	0.194
4	180.2	0.248
tot	723.8	0.998
av	181.0	
x area	3.922	Amps

Figure 8.3
Current densities, Core 1

We would expect that product to be the 4.0 Amps we applied to the trace in the simulation. This result, which is typical of all the simulations to follow, represents about a 2 % difference from what is expected. This is because we are approximating an integration with only four points. See the "Caution" in Appendix 6.1.

As can be seen from Appendix 6.2, as we go from core layer 1 to core layer 4, current densities decrease at the front of the via and increase at the rear of the via. At core layer 4, the mid-pint of the via, the current density is uniform around the via wall (at 181.3 A/mm^2).

TRM simulations are "steady state" (think DC) in nature. We normally think of current density being uniform in a conductor in steady state [1]. So why are there gradients in the via wall? We believe the answer relates to the length of the current path. The *shortest* current path starts at the front (leading edge) of the via, wraps around each side of the via as you progress downwards through the via, and exits the via at the leading edge of the via *with respect to the bottom trace* (which is at the far side of the top trace.) All other paths are longer by comparison. Therefore, the current through the via divides inversely proportional to the path length.

The current density on the top (trace) layer, Figure 8.4, results in a very interesting pattern. The density is higher at the front edge and lower at the back edge, which might be expected. But the current density totals to the equivalent of almost 4.4 Amps (instead of the 3.92 Amps measured in the via itself.) That is because not all the current at the front edge of the via actually enters the via at the front edge! Part of it continues past the front edge and then enters the via, for example, at the sides. If we think of current density as representing tiny "packets" of current, some of the packets get counted two (or perhaps even more) times. (This effect will be more pronounced when we look at the case of multiple vias.)

When looking at Figure 8.2 it is tempting to compare the current patterns on the trace to water flowing along a trace and then into a hole in the trace. Now I am well aware of the negative attitudes most electrical engineers have towards hydraulic analogs of current flow. Nevertheless, the resemblance is pretty striking.

When thinking of current density as little packets of current that may get counted multiple times on the trace layer, this effect extends a small distance down into the via itself. We ran another simulation that, in effect, separated Core 1 into three individual cores with thicknesses 15, 15, and 200 um respectively (approximating the 230 um of Core 1.). Recall that the trace thickness on the top layer is 35 um. These results are summarized in Appendix 6.3. Note how there is still a tendency for the total current (as calculated by multiplying average current density by area) to exceed 3.92 Amps even in the second 15 um layer. After that, the densities are about the same as they were in the previous simulation.

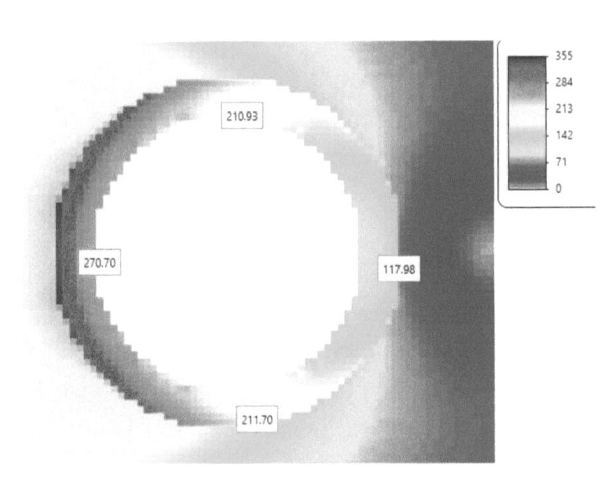

	density	Fraction of total
1	270.7	0.373
2	210.9	0.291
3	118.0	0.163
4	211.7	0.292
tot	811.3	1.119
av	202.8	
x area	4.397	Amps

Figure 8.4
Current densities on the top layer.

8.2 Multiple Vias

We then simulated a case with four vias, arranged in the pattern shown in Figure 8.5. In this simulation the trace width is 0.9 mm (35 mil), and still 34 um thick. The current in this simulation was still 4.0 Amps.

The current density patterns of the first two layers (trace layer and Core 1) are shown in Figure 8.6. The full set of current density patterns by layer is provided in Appendix 6.4. The patterns are symmetrical around the horizontal center line. Therefore, we can take the current density readings for either the upper via (in the figure) or lower via; the other via will be the same.

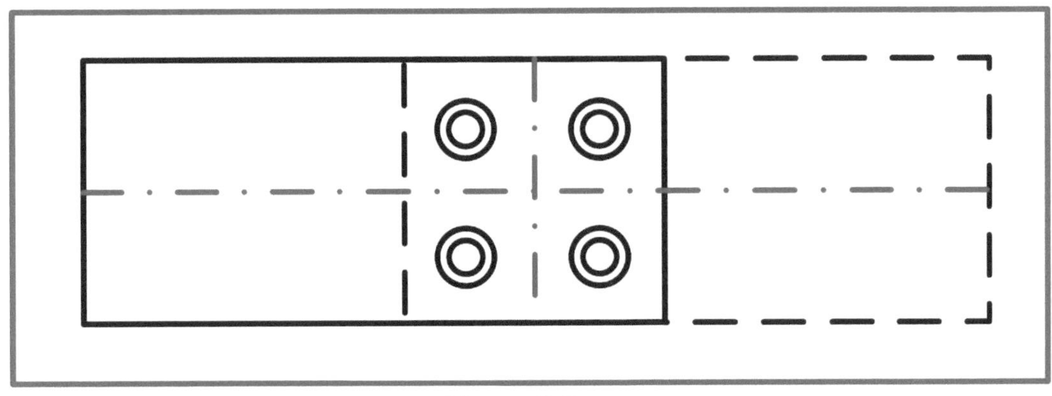

Figure 8.5
Simulation of multiple vias (not to scale)

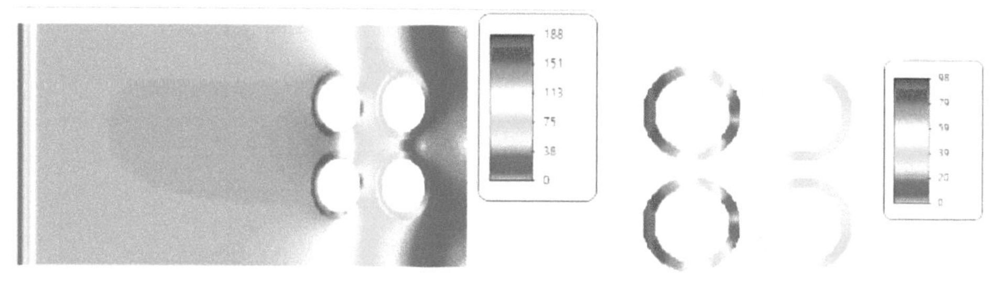

Figure 8.6
Current density for trace layer (left) and Core 1 layer (right).

We would expect the current to divide evenly among the four vias, Looking at the current densities in the various core layers confirms that this is true. Each via is carrying (approximately) 0.98 Amps. The total calculated current is 3.94 Amps (compared to 4.0 Amps modeled and the 3.92 calculated with the single via case.) (Again, the total probably does not equal 1.0 Amps each because we are not integrating the area. Again, see the "Caution" in Appendix 6.1.) The current pattern through each via wall looks very similar to that for the single via case. The density is highest at the "front" of the via and lowest at the back of the via on the outside layers, and uniform around the via wall at the mid point. That is because the shortest path length through the via is as before; enter the via at the front, travel around the barrel as we go down the via, and out the "front" with respect to the bottom trace layer. The total current path through each via is identical for each of the four vias.

But the current densities on the top (trace) layer are more interesting than before (see Figure 8.7). The total current around the front via (found by multiplying the sum of the densities by the via conducting area) is almost 2.3 Amps (instead of the 1.0 Amp flowing through the via.) As before, this is because some of the current seen at that point is flowing into the via, but some is flowing past the front via to the rear via. Using the analogy of little current packets, the incidence of "double counting" current packets is much greater in this simulation. The data in Appendix 6.4 suggests that this tendency is still slightly apparent in the front vias in the first core layer where the "total" calculates to 1.1 Amps and then 0.98 Amps in succeeding layers.

		Fraction				Fraction
	density	of total			density	of total
1	136.87	0.753689		1	76.4	0.420705
2	129.2	0.711454		2	45.1	0.248348
3	33.77	0.185958		3	29.6	0.162996
4	124	0.682819		4	66.8	0.367841
tot	423.84	2.333921		tot	217.9	1.19989
av	105.96			av	54.475	
x area	2.299332	Amps		x area	1.182108	Amps

Figure 8.7
Current density data on the top (trace) layer for the four via simulation. The front vias are on the left, the rear vias on the right.

8.3 Multiple Vias and Turn

In our final simulation we used the same trace dimensions as the previous simulation, but we turned the bottom trace 90 degrees (see Figure 8.8).

The current distribution in the trace layer and in the first core layer are shown in Figure 8.9. The full pattern of current densities through all the layers is provided in Appendix 6.5.

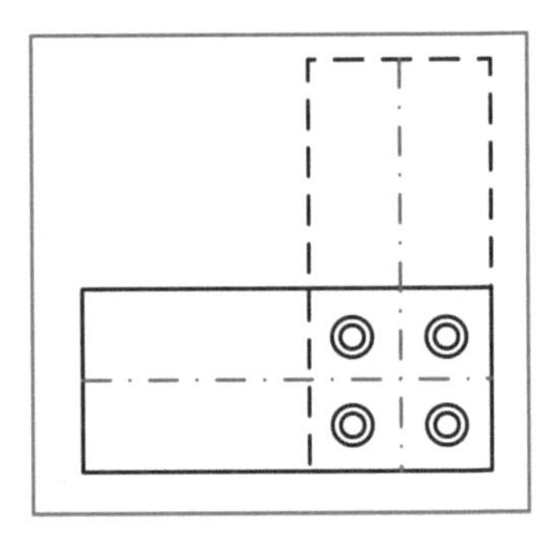

Figure 8.8
Simulation of four vias with traces turned 90 degrees.
(Not to scale)

By now we can almost predict what will happen. The shortest current path will be through the upper left (in Figure 8.9) via, and the longest path through the lower right via. So we would expect the current to divide unequally through this configuration. In fact it does, with the upper left via carrying 1.14 Amps and the lower right via carrying 0.72 Amps. The other two vias carry 0.96 Amps. Beyond that, the current densities in the via, as we progress down through the layers, are pretty much as in the prior simulation.

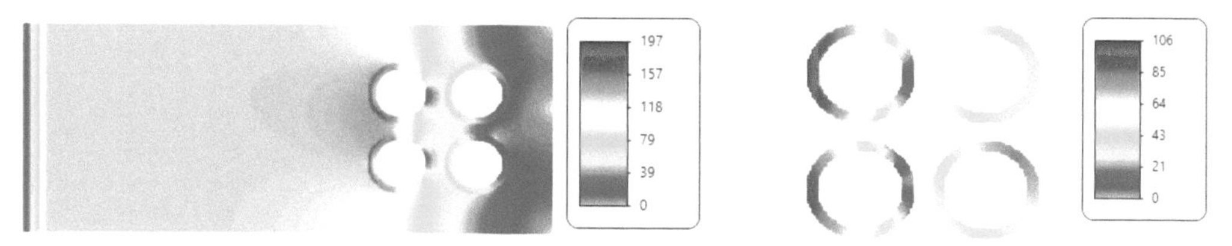

.Figure 8.9
Current density for trace layer (left) and Core 1 layer (right).

8.4 Conclusions

I offer the following thoughts about the foregoing:

1. The current density gradients do not contribute in any way to trace and via temperature. The thermal conductivity of all the copper in the area distributes the heat away from any potential point heat generation.
2. TRM results in a steady-state (DC) analysis. I am not aware of any high-speed signal integrity inferences we can make from these observations.
3. Frankly, while these results are interesting, I can't think of any particularly practical import to them! The current density variations are small enough that the localized voltage variations are minimal, and there are no other areas I can think of where these variations might pose a practical problem. Johannes and I leave it to others to explore this thought.

Notes:

1. Non-uniform current densities are common, for example, when dealing with higher frequencies --- the skin effect is one familiar situation.

9 FUSING CURRENTS: BACKGROUND

At the end of a seminar one day, a participant asked me the following question:

> If there is a catastrophic failure of my system, a certain 1 Oz. trace will carry 40 Amps. I need it to carry that for 1 second while the system shuts down in a controlled manner. What size trace do I need?

The first time I was asked this question I referred the person to people in the fusing industry. The second time I was asked I went to the people in the fusing industry myself and found that they didn't have a clue how to proceed!

It turns out the early work on this topic goes back to W. H. Preece in the 1880's, and I. M. Onderdonk, possibly in the 1920's. Preece's work has been well documented, although until recently copies of that documentation have been difficult to obtain. There is growing speculation, however, that Onderdonk may never have published his source data under his/her own name.

In this chapter I will (a) discuss the background of Preece, his motivating factors, and links to his source data; (b) discuss what can be inferred about Onderdonk, his likely motivating factors, and links to early work based on his equation; (c) outline the critical assumptions behind these formulas, and (d) in the absence of any Onderdonk source data, independently derive Onderdonk's equation.

9.1 W. H. Preece

In the 1880's, Sir William Henry Preece was a consulting engineer for the British General Post Office, and became Engineer-In-Chief in 1892 [1]. At that time, the Post Office was responsible for the telegraph (and later wireless telegraph) system in England. He published three papers [2] in the Proceedings of the Royal Society of London in the 1880's that formed the basis for his famous equation:

[Eq. 9.1] $I = a * d^{3/2}$

where d is the diameter of the wire in inches, a is a constant (10244 for copper), and I is the fusing current in Amps. A little algebra transforms this equation to:

[Eq. 9.2] $I = 12277*A^{3/4}$

where A is the cross-sectional area in inches^2.

Interestingly, in the 1890's Preece became an ardent supporter, both politically and financially, of Guglielmo Marconi and his work on trans-oceanic wireless telegraphy.

Preece was motivated by the problem of lightning striking telegraph wires. The early form of lightning arrester used in that period is shown in Figure 9.1, reproduced from Preece's 1883 paper. Most of the energy from the strike was intended to flow through the capacitor, but some was left to travel onto the "cable," which led to the telegraphy equipment. Preece wanted to determine the best material, and its size, for the lead from the arrester to the cable. The wire should be able to carry the normal current for telegraph operations, but fuse at currents slightly above that level.

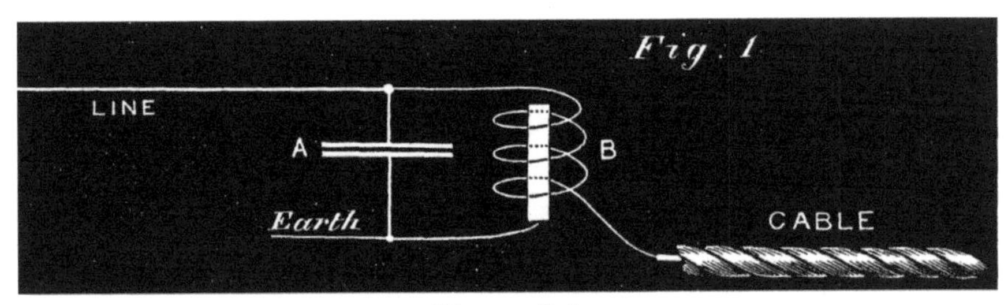

Figure 9.1
Early form of lightning arrester. Source: Preece's 1883 paper.

Preece did not approach this problem analytically. He was, what we might call today, a "lab rat" and did all his experiments in the lab. He would apply current to a test wire until the wire began to glow. He would call that point the "fusing" point (more appropriately the "melting" point). He looked at a large selection of materials and provided the constant (a in Equation 9.1) for each material. He concluded that "the best metal to use for small diameters was platinum, and for large wires tin" [1887 paper]. The final constants he derived were summarized in his 1888 paper, which is where the value 10244 is found.

9.2 I. M. Onderdonk

Onderdonk, however, is much more obscure. Speculation is growing that Onderdonk never published his work under his (or her?) own name, and we have uncovered no citations for any work by Onderdonk himself. His equation is offered almost as a given, in much the same way we reference Ohm's Law today. His equation appears in several papers and sources [3 through 7], usually in the form

[Eq. 9.3] $$33\left(\frac{I}{A}\right)^2 S = \log_{10}\left(\frac{t}{274}+1\right)$$

for T = 40 °C, but occasionally in the more general form:

[Eq. 9.4] $$33\left(\frac{I}{A}\right)^2 S = \log_{10}\left(\frac{t}{234+Ta}+1\right)$$

Where: I = the current in Amps
A = the cross-sectional area in circular mils [8]
S = the time in seconds the current is applied
t = the rise in temperature from the ambient or initial state [9]
Ta = the reference temperature in degrees C

Some publications erroneously cite Onderdonk's name in a citation. I will repeat two of those citations here, anonymously to protect the authors:

Onderdonk, J. M., Short-Time Current Required to Melt Copper Conductors, Electrical World ***121****:26, 98 (24 June 1944).*

I. M. Onderdonk and E. R. Stauffacher, "Short Time Current Carrying Capacity of Copper Wires", G.E.Review, pp.326 1928

The first [4] is an anonymous (not by Onderdonk, as the citation suggests) single page nomograph inserted in the "Electrical World" magazine. It is provided in Appendix 7. There are notes on the face of the nomograph that say (somewhat cryptically, and among other things) "...based on an equation by I. M. Onderdonk," and "Chart from engineering department of a midwestern utility."

The second [5] is a respectable article written by E. R. Stauffacher (but NOT co -authored by Onderdonk). It contains a different nomograph (reproduced here as Figure 9.2). Stauffacher comments that the "Values of short-time current required to melt copper conductors, as given by the chart shown in Fig. 1, were calculated from the formula..." and then he offers Onderdonk's equation. A footnote to the equation says "This formula was developed by I. M. Onderdonk."

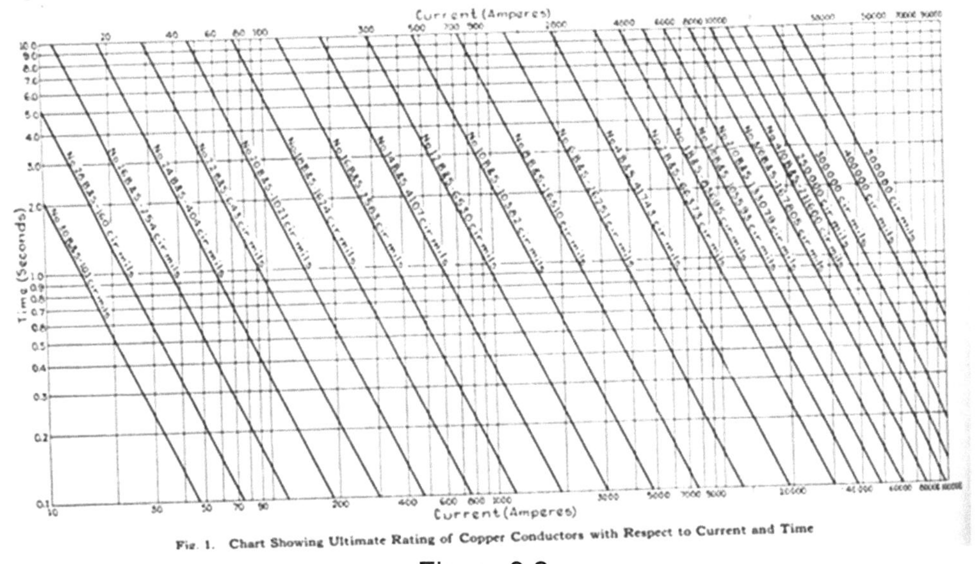

Fig. 1. Chart Showing Ultimate Rating of Copper Conductors with Respect to Current and Time

Figure 9.2
Nomograph from E. R. Stauffacher, Reference [5].

It is tempting to speculate that, since Stauffacher is the earliest reference Johannes and I have found, Onderdonk may have been a contemporary of Stauffacher. Stauffacher, himself, was employed as Superintendent of Protection at Southern California Edison when he wrote his article (in 1928). Perhaps Onderdonk was an engineer at SCE at the same time.

Stauffacher's article offers a clue as to what the motivation was for Onderdonk's equation. In approximately that same time frame, SCE was building a high power transmission line from Southern California to the construction site for Boulder Dam (now Hoover Dam) in Nevada. The concern was that under certain conditions there may be a short circuit that occurs between a high-voltage transmission line and the copper wires attached to the insulators or poles supporting the line. These copper wires needed to carry that short-circuit current for the short period of time "required to clear the faulty line from the system." Thus, the desirability of having a variable for time (S in the equation).

9.2.1: Cautions: It is important to understand what constitutes the "melting" of a copper wire. When we apply heat to a wire (or any solid for that matter), there are two stages we can consider. The first is that the wire heats up from the initial temperature to the melting temperature. Let's call this time t1. Then additional heat (and time, say time t2) is required to change the wire from a solid to a liquid (i.e. to actually melt the wire.) This second amount of heat is referred to as the heat of fusion [10]. Time t1 is required to heat the wire TO the melting temperature, and time t1 + t2 is required to actually melt the wire from the initial temperature.

Even if the trace is beginning to melt, current may flow through the liquid copper. However, liquid copper has a lower electrical and thermal conductivity than does solid copper (refer back to Table 3.1 and Figure 3.3). Therefore, once the liquid starts to form, there may be an "explosive" run-away condition that follows. The circuit will "break" (i.e. current will stop flowing) only if the liquid path separates. This may happen as a result of gravity, as a result of constricting surface tension, or as a result of explosive "splattering." Therefore, if we consider time t2 to be the time the circuit path opens, that can depend on many subtle variables.

Preece clearly refers to time t1. He raised the current to the point where the wire glowed. Onderdonk's equation also only applies to time t1. This is inferred from most (but not all) of the references and is confirmed by the derivation in the next section.

Unfortunately the semantics found in some of the articles seems to confuse this point. For example, the nomograph in Electrical World is titled "Short-Time Current Required to ***Melt*** Copper Conductors" (emphasis added.) Stauffacher refers to the "short-time current required to melt copper conductors." But since both these are based on Onderdonk's equation, they actually only refer to the time to bring the temperature TO the melting point.

Preece's equation is based on slowly bringing the current up to the melting point. But he did not bring the variable of time into the equation. There would have been some cooling effects that occurred while he was heating the various materials, but this was of little consequence in what he was trying to discover.

Onderdonk, on the other hand, specifically brought time into the relationship. Thus he had to make an assumption about the cooling effects that might have coexisted with the heating effects. His assumption was that there were no cooling effects (from conduction, convection, or radiation.) Thus, the heating was rapid (as might happen in a short-circuit situation.) That is why the references say that Onderdonk's equation is only valid for short periods of time, typically less than 10 seconds. People can quibble about this time. For example, Adam [11] suggests that cooling effects can begin to affect the results in as short as one or two seconds of time. In any event, Onderdonk's equation becomes less and less reliable as time increases.

9.3 Derivation of Onderdonk Equation

(I am indebted to Johannes Adam for this derivation. What follows is modified slightly from his original derivation, which can be found in [11])

Johannes and I are aware of only one serious attempt at independently deriving Onderdonk's Equation (there might well have been other attempts we are unaware of.) Babrauskas and Wichman [6] did so with interesting results. Their equation 6 derived the time to heat a copper wire to the melting temperature. Their equation 7 derived the time to melt the wire. Their equation 8 was the sum, which they then compared to Onderdonk's equation and remarked that their result was 17% higher than Onderdonk's. We believe that that comparison is not appropriate and they should have compared their equation 6 with Onderdonk's. Our comparison of their equation 6 with Onderdonk's suggests that their equation is just 4% below Onderdonk's, acceptably close.

Table 9.1 includes a list of variables used in the derivation:

9.3.1 Basic equation: The following equation (Equation 9.5) describes an equality in a piece of matter over a time span Δt (starting at t=0) [12].

Gain of thermal energy (l.h.s.) = Joule heating by electric energy (r.h.s.)

[Eq. 9.5] $$c_p \cdot M \cdot \Delta T = R \cdot I^2 \cdot \Delta t$$

Because there is no heat loss over the surface, temperature would rise infinitely with time. The units on both sides are W×s=Joule. M and R depend on the geometry of the wire, c_p is a material value.

I	(DC) current	Ampere	
t	time	sec	
T	wire temperature (avg.)	deg C	
Θ	Temperature rise $T–T_{ref}$	Degree C	
L	Length of wire	meter	
M	Mass of wire	kg	
R	Electric resistance of the wire	Ohm	
c_p	Specific heat capacity (copper)	J/(kg*K)	385
ρ	density of wire material (copper)	kg/m^3	8900
ρ_{20}	electric resistivity @20 °C (copper)	$Ohm*m^2/m$	1.72×10^{-8}
α	Temperature coefficient of resistivity	1/° C	.00393@20°
A	Cross-sectional Area	m^2	
K	Temporary utility variable	-	

Table 9.1
Variables used in the derivation

The electric resistance of a wire is temperature dependent. If the ambient temperature is 20° C, the resistance at any other temperature, T, is typically expressed as

$$R = \rho_{20}(1+\alpha_{20}\cdot(T-20))\frac{L}{A}$$

[Eq. 9.6]

Using the basic relation for volume and mass we rewrite Equation 9.6 as:

$$c_p\cdot\rho\cdot L\cdot A\cdot\Delta T = \rho_{20}(1+\alpha_{20}(T-20))\frac{L}{A}\cdot I^2\cdot\Delta t$$

[Eq. 9.7]

The length of the wire will cancel. If we assume the initial temperature of the wire also being 20 degC (see again note 9), we can unify ΔT and T-20 to a temperature rise Θ=T-20. Because of the permanent link between t and Θ, Θ (t) will be time-dependent and the correct basic equation is an ordinary differential equation:

$$c_p\cdot\rho\cdot\frac{d\Theta}{dt} = \rho_{20}(1+\alpha_{20}\Theta(t))\frac{I^2}{A^2}$$

[Eq. 9.8]

Note that I/A is current density (A/m^2).

9.3.2 Solving the Equation: We solve Equation 9.8 step by step. First we collect the parameters in an auxiliary quantity K

$$K = \frac{\rho_{20}}{c_p \cdot \rho} \frac{I^2}{A^2}$$

which leads to

[Eq. 9.9] $$\frac{\mathrm{d}\Theta}{\mathrm{d}t} = K(1 + \alpha_{20}\Theta)$$.

Then we substitute

$$\vartheta := 1 + \alpha_{20}\Theta \rightarrow \frac{d\Theta}{dt} = \frac{d\Theta}{d\vartheta}\frac{d\vartheta}{dt} = \frac{1}{\alpha_{20}}\frac{d\vartheta}{dt}$$

and get a simple ordinary differential equation,

[Eq. 9.10] $$\frac{\mathrm{d}\vartheta}{\mathrm{d}t} = \alpha_{20} K \vartheta$$.

Initial condition: at time t=0 the wire shall be at 20 degC:

$\Theta(t=0)=0 \rightarrow \vartheta(t=0)=1$

The way to solve Equation 9.10 is to integrate:

$$\int_1^{\vartheta} \frac{\mathrm{d}\vartheta'}{\vartheta'} = \int_0^t \alpha_{20} K \mathrm{d}t'$$

which gives Equation 9.11 [13]

[Eq. 9.11] $$\ln(\vartheta) - \ln(1) = \alpha_{20} K(t - 0)$$

or (because ln(1)=0)

[Eq. 9.12] $$\ln(\vartheta) = \alpha_{20} K t$$

We then unfold the aux variables

[Eq. 9.13] $$\ln(1+\alpha_{20}\Theta) = \frac{\alpha_{20}\rho_{20}}{c_p \cdot \rho}\frac{I^2}{A^2}\cdot t$$

and exchange rhs and lhs (and move α to the denominator, for reasons that will be apparent in a moment)

[Eq. 9.14] $$\frac{\alpha_{20}\rho_{20}}{c_p \cdot \rho}\frac{I^2}{A^2}\cdot t = \ln\left(1+\frac{\Theta}{1/\alpha_{20}}\right)$$

We are going to solve the l.h.s of this equation separately. First, looking at

[Eq. 9.15] $$\frac{\alpha_{20}\rho_{20}}{c_p \cdot \rho}\frac{I^2}{A^2}$$

we rely on the relationship that $\alpha_{T2} * \rho_{T2} = \rho_{T1} * \alpha_{T1}$ at any two reference temperatures (for a given material.) (See proof in [14]). Therefore, we can use the 20° C reference values for ρ and α regardless of the reference temperature for the equation. Note also that typical expressions for Onderdonk's Equation 9.3 and 9.4 are in units of circular mils. So:

Our material data of copper are:

- ρ_{20} = 0.0172 Ω mm²/m (at 20 degC) or 1.75×10^{-8} Ω m²/m
- α_{20} = 0.00393 1/K, (at 20 deg C)
- c_p = 385 J/kgK
- ρ = 8900 kg/m³.

To calculate the coefficient (equation 9.15) in front of $(I/A)^2$ let us use our material data and convert equation 9.14 to circular-mils

SI Units : $\frac{\alpha_{20} \cdot \rho_{20}}{c_p \cdot \rho}$ $= 0.00393 \cdot 1.72 \cdot 10^{-8} /(385 \cdot 8900) = 1.973 \cdot 10^{-17}$
1 m² =1.98 10^9 circ-mils
We have to compensate A² → 1 $m^4 = 3.9 \cdot 10^{18}$ circ-mils²
Introducing to a) → $1.973 \cdot 10^{-17} \cdot 3.9 \cdot 10^{18} = 76.9$
From ln to $\log_{10}$ we have to divide by ln(10) = 2.30 → 77/2.30 = 33.5

So, equation 9.14, with *A* in circular mils, is:

$$33.5 \left(\frac{I}{A}\right)^2 \cdot t = \log_{10}\left(1 + \frac{\Theta}{1/\alpha_{20}}\right)$$

[Eq. 9.16]

Now, to solve for 1/α for any reference temperature:

We note that $\alpha_{T2} * \rho_{T2} = \rho_{T1} * \alpha_{T1}$

This leads to

$$1/\alpha_{T2} = \rho_{T2} / \rho_{T1} * \alpha_{T1} = (\rho_{T1} + \rho_{T1} * \alpha_{T1}(T2 - T1)) / \rho_{T1} * \alpha_{T1}$$

Or:

$$1/\alpha_{T2} = \frac{\rho_{T1}}{\rho_{T1} * \alpha_{T1}} + (T2 - T1) = \frac{1}{\alpha_{T1}} + (T2 - T1)$$

If we let T1 = ambient = 20 °C and T2 be any other reference temperature, T_{ref}, then

$$1/\alpha_{ref} = \frac{1}{\alpha_{20}} + T_{ref} - 20 = 254 - 20 + T_{ref} = 234 + T_{ref}$$

Leading to our final expression

$$33.5 \left(\frac{I}{A}\right)^2 \cdot t = \log_{10}\left(\frac{\Theta}{234 + T_{ref}} + 1\right)$$

[Eq. 9.17]

Which at a reference temperature of 40°C (Onderdonk's reference) is

[Fq. 9.18] $$33.5\left(\frac{I}{A}\right)^2\cdot t=\log_{10}\left(\frac{\Theta}{274}+1\right)$$

or equal to equations 9.3 and 9.4.

It is sometimes desirable to have equation 9.18 expressed in terms of t, the time to reach the melting temperature. Solving for t leads to

[Eq. 9.19] $$t=\left(\frac{1}{33.5}\right)\left[\log_{10}\left(\frac{\Theta}{234+T_{ref}}+1\right)*\left(\frac{A}{I}\right)^2\right]$$

where A is still in circular mils.

Using Equation 9.19, we can now transform Onderdonk's Equation to any temperature reference point. I offer it with reference temperatures of 20° C, 40° C, 85° C, and 105° C without further calculations. The latter temperature references are important in the automotive industries. (Note that the higher the reference temperature the shorter the time is to reach the melting temperature.)

Reference Temp °C	Denominator in Log expression
20	254
40	274
85	319
105	339

Using the melting point of copper (1083 °C) and setting T_{ref} = 20 °C, Equation 9.19 can be reduced to the form:

[Eq. 9.20] $t = c*(A/I)^2$

where c = .0213 in circular mils.

The values for c are supplied in in the following table for various units and reference temperatures.

Reference Temp	C in circular mils	C in square mils	C in square mm
20	.0213	.0346	8.30 x 10^4
40	.0203	.0330	7.92
85	.0184	.0298	7.15
105	.0176	.0285	6.85

Notes:

1 http://en.wikipedia.org/wiki/William_Henry_Preece

2 Preece W. H., On the Heating Effects of Electric Currents, Proc. Royal Society 36, 464-471 (1883). No. II, 43, 280-295 (1887). No. III, 44, 109-111 (1888).These documents have been made available on-line in recent years (you may need to copy-paste to make these links work):

http://rspl.royalsocietypublishing.org/content/36/228-231/464.full.pdf+html
http://rspl.royalsocietypublishing.org/content/43/258-265/280.full.pdf+html
http://rspl.royalsocietypublishing.org/content/44/266-272/109.full.pdf+html

or at www.ultracad.com/articles/reprints/preece.zip

The following five sources (among others) make reference to Onderdonk's equation:

3 "Standard Handbook for Electrical Engineers," 12 Ed., 1999, McGraw-Hill, p. 4-74

4 Anon., "Short-Time Current Required to Melt Copper Conductors," *Electrical World* **121**:26, 98 (24 June 1944). (Download a copy from www.ultracad.com/articles/reprints/electrical_world.pdf)

5 E. R. Stauffacher, "Short-time Current Carrying Capacity of Copper Wire," General Electric Review, Vol 31, No 6, June 1928 (Download a copy from www.ultracad.com/articles/reprints/stauffacher.pdf)

6 Vytenis Babrauskas and Indrek Wichman, "Fusing of Wires by Electrical Current," Fire and Materials 2011 Conference Proceedings, Interscience Communications (download a copy from https://www.academia.edu/9357019/Fusing_of_Wires_by_Electrical_Current

or from www.ultracad.com/articles/reprints/babrauskas.pdf

7 Norocel Codreanu, Radu Bunea, and Paul Svasta, "New Methods of Testing PCB Traces Capacity and Fusing," Note: Their copy of Onderdonk's formula is incorrect and contains a misprint! (Download a copy from www.ultracad.com/articles/reprints/codreanu.pdf)

8 A circular mil is the area of a circle with a diameter of one mil. The formula is $A = d^2$. The conversion from circular mils to mil^2 is $\pi/4$. Normal conversions are:

$1\ mil^2 = 1.273$ circular mils

1 circular mil = $.7854\ mil^2$

$1\ m^2 = 19.736*10^8$ circular mil

1 circular mil = $5.067*10^{-10}\ m^2 = 5.067*10^{-4}\ mm^2$

9 From a thermodynamic engineering standpoint, there is a significant difference between the ambient temperature and the initial condition. For example, there could be some heating of a wire above the ambient temperature in a normal operation before a change of current is applied. In this paper we are assuming that the initial condition (temperature) is the same as the ambient temperature.

10 See http://en.wikipedia.org/wiki/Melting_point

11 Johannes Adam white paper No.10

www.adam-research.de/pdfs/TRM_WhitePaper10_AdiabaticWire.pdf

12 l.h.s means left hand side of the equation, r.h.s mean right hand side.

13 For those of us who have forgotten,

$$\int \frac{dx}{x} = Ln(x)$$

14 See Appendix 3 in Johannes Adam, Douglas Brooks, *In Search For Preece and Onderdonk* available at www.ultracad.com .

10 FUSING CURRENTS: ANALYSES

As I mentioned at the beginning of Chapter 9, at the end of a seminar I was leading one day, a participant asked me the following question:

> If there is a catastrophic failure of my system, a certain 1 Oz. trace will carry 40 Amps. I need it to carry that for 1 second while the system shuts down in a controlled manner. What size trace do I need?
>
> (See supplemental comment at end of chapter)

In Chapter 9 I covered the historical background investigations into this topic. Of relevance to us is Onderdonk's Equation, repeated here as Equation 10.1.

$$33\left(\frac{I}{A}\right)^2 S = \log_{10}\left(\frac{\Delta T}{234+Ta}+1\right)$$

[Eq. 10.1]

A more useful form of Onderdonk's equation (at least for us) is Equation 10.2, in which we rearrange the terms to solve for t (time, which is S in Equation 10.1.)

$$t=\left(\frac{1}{33.5}\right)\left[\log_{10}\left(\frac{\Delta T}{234+T_{ref}}+1\right)*\left(\frac{A}{I}\right)^2\right]$$

[Eq. 10.2]

The various variables are:

- I = the current in Amps
- A = the cross-sectional area in circular mils [1]
- S = t = the time in seconds the current is applied
- ΔT = the rise in temperature from the ambient or initial state [2]
- Ta = the reference temperature in degrees C

In this chapter we are going to look at the question of trace fusing currents using Onderdonk's equation and also some thermal simulation models of traces using the same techniques described in Chapter 6.

10.1 "Fusing" Time and Temperature

It is important to understand what we mean by fusing temperature. I covered this in detail in Chapter 9 and I refer you back to Section 9.2.1. Fusing temperature in our case is the melting temperature of copper, 1083° C. Fusing time is the time to reach the fusing temperature (but not the additional time necessary to actually melt the trace.

10.2 Assumptions and Cautions

In this chapter we are talking about the fusing temperature of PCB traces. We are not talking about fuses *per se*. So, for example, we are not talking about a commercial fuse of the type shown in Figure 10.1(a) nor are we talking about forming a fuse link along a PCB trace of the type shown in Figure 10.1(b) (although the principles we explore here probably apply equally well to Figure 10.1(b).) Nor are we talking about an insulated wire. Instead, we are talking about standard PCB traces of uniform width and uniform thickness over their entire length.

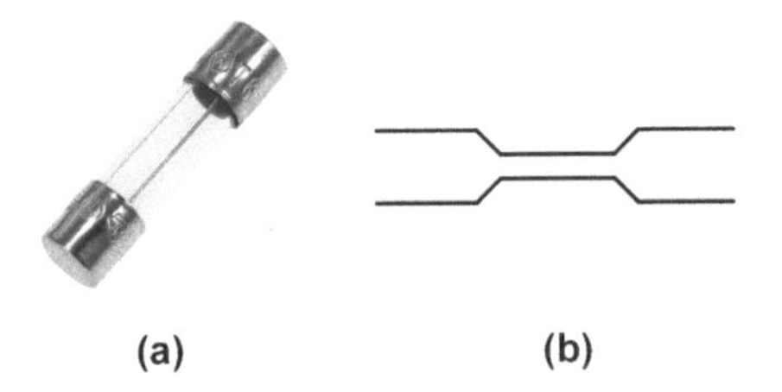

(a) (b)

Figure 10.1
Types of fuses we are not considering

10.3 Simulation Models

In Chapter 6 we developed a thermal simulation model of a 6.0" trace and used that in various sensitivity scenarios. We will continue to use a similar simulation model in this chapter. This model is referred to in our graphs as "TRM Trace". The various forms of this model range in thickness from 1 Oz. to 2 Oz., and in width from 20 mil to 200 mil. The trace model is placed on a 63 mil thick FR4 substrate.

I also developed a new model for this chapter I call "TRM Fuse." It is a solid strip of copper, 200 mil long. It also varies in width from 20 to 200 mil and in thickness from 1 to 2 Oz. There is no substrate under this trace. You might envision it as a trace segment placed over a hole drilled in the board substrate material. It is illustrated in Figure 10.2

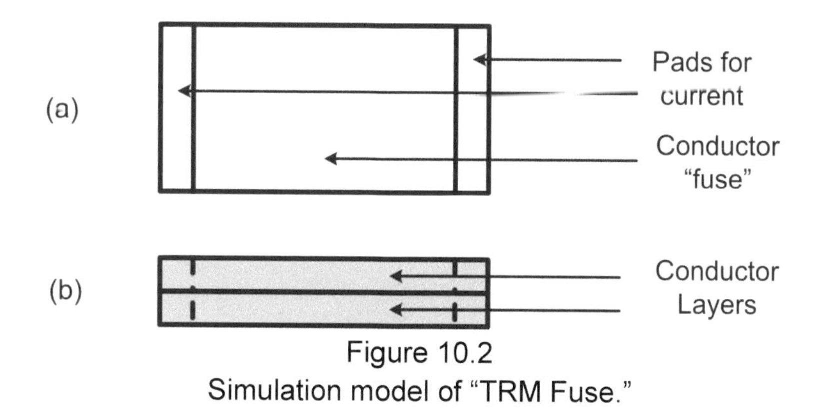

Figure 10.2
Simulation model of "TRM Fuse."

Since TRM is not designed to model a single layer structure, I modeled the "fuse" on two layers as shown in Figure 10.2. Both conductor layers are copper; there is no dielectric layer. The Heat Exchange Coefficient (see Chapter 6) was set to zero, but setting it to 10 had negligible effect on the simulation model results because the time frames are so short.

There is a subtle detail involved in running a simulation of this type. In order to get a precise measure of fusing time, the model needs to run for a period of time, calculate the temperature of the copper, adjust the resistivity of the copper to the new temperature, and then run the simulation for the next period of time. This continues until the melting point of copper (1083 degrees C) is reached. Each "period of time" is called a "Step." The step size impacts the results; the most precise results are reached with infinitely small steps (i.e. in the limit where step approaches zero.) A step size that small results in very long calculation times. After considerable experimentation we found that setting the step to 0.1 seconds was a reasonable compromise.

10.4 Simulation Results

I have expressed Onderdonk's Equation in the form for solving for time in Equation 10.2. In that equation, A is still in units of circular mils. If we assume a reference temperature of 20 °C, a fusing temperature of 1083° C, and convert this to square mils, the result is Equation 10.3:

[Eq. 10.3] $t = .0346*(A/I)^2$

We can plot this equation as a curve and it represents the theoretical fusing time for a conductor (at least as based on Onderdonk's Equation.) Then if we run a simulation at various currents and plot the results of the simulation, we can compare the simulation model results with the theory. Figure 10.3 is the result for 1.0 Oz. 100 mil wide trace.

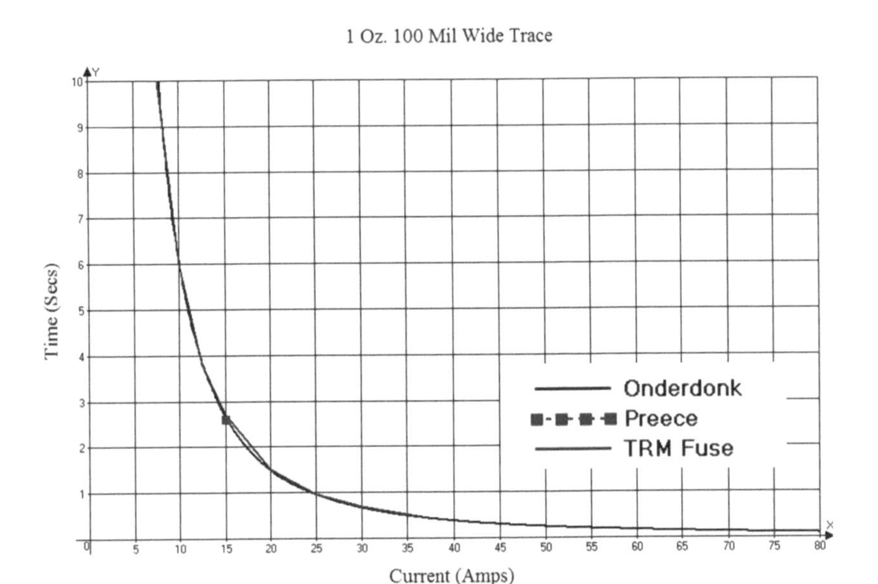

Figure 10.3
Simulation results against Onderdonk's Equation.

The simulated results lie practically on top of the theoretical results! The fit could not be better. This gives us a reasonably comfortable feeling that our approach here is viable. The red dot on the curve is the implied point from Preece's equation (see Chapter 9) for the same level of current.

Six different simulations were run, 1 Oz.(1.35 mil) and 2 Oz. (2.7 mil) thicknesses with widths of 20 mil, 100 mil, and 200 mil. The results for all other configurations are identical to this one except for scale. The results for all configurations are provided in Appendix 8. This gives us a high degree of confidence in TRM's ability to simulate a fuse in this kind of simulation.

I constructed a trace model for simulation that was a very simple PCB trace, 155 mm (approximately 6.0") long. It varied in width and thickness depending on what I was testing. There were 3 mm (118 mils) between each side of the trace and the edge of the board, and the ends of the trace were 5 mm (200 mils) in from each edge. A schematic of one of the traces is shown in Figure 10.4, which is a figure from one of the TRM simulations.

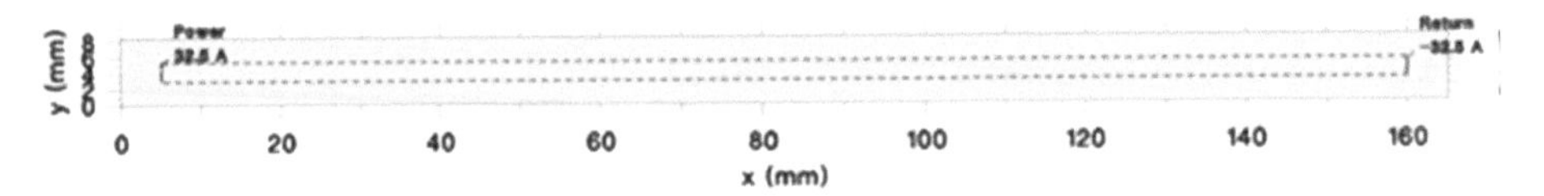

Figure 10.4
Schematic of our trace model for simulation.

The same six configurations as described above were run on this trace model. The 1 Oz. 100 mil wide simulation result is shown in Figure 10.5. (The results for all six configurations are shown in Appendix 8.)

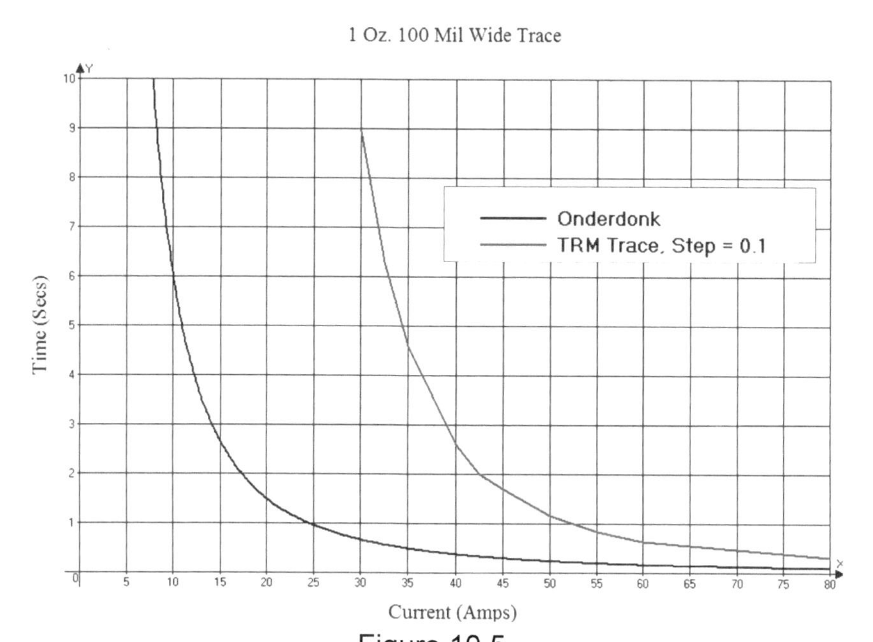

Figure 10.5
Fusing time vs. current for trace simulation, compared to Onderdonk's equation.

There are three types of cooling that occurs when these traces heat --- conduction into the board material, convection into the air, and radiation away from the trace. The Heat Exchange Coefficient was set in the simulations to 10 to represent a normal situation. Subsequent testing showed that the value of the Heat Exchange Coefficient did not change the results significantly, showing that the primary cooling effect in the time frames looked at here are confined to conduction through the board material. The significant shift of the trace model simulation to the right of Onderdonk's equation is the result of conduction of heat into the board material, thereby slowing the increase in temperature.

The dielectric material used in this simulation was FR4. Polyimide might provide slightly better heat conductivity, better cooling, and therefore slightly longer fusing times. The size of the board had no effect on the results. Placing a plane on the opposite side of the board did not affect the results. Placing a plane 12 mils under the trace layer did have an effect on the results, increasing the fusing time by 30% to 100 % depending on the current level (lower current levels, and therefore longer fusing times, resulted in higher percentage changes).

An interesting observation of these results is that at relatively high currents (and therefore at relatively short times) the curves begin to merge. This is the area where the cooling effects of the board dielectric material have not had time to “kick in.”

I also ran a simulation on the trace model we used for simulating the IPC test board (see Chapter 6.) The results were *identical* with those from our trace model (within less that 0.1 second at every current level.) This gives further confidence in our feeling that the fusing time is a function of trace size and board material only, and nothing else (except a closely spaced underlying plane.) This also gives further confidence in TRM’s ability to model fusing currents and times.

10.5 Short-time Effects

Let’s look back at Equation 10.3 and rearrange terms:

[Eq. 10.4] $t\,(I/A)^2 = .0346$

This says that the product on the left is a constant [3]. Figure 10.6 plots the result of $t^*(I/A)^2$ for the various fuse model simulations. The results are the same as the Onderdonk result within reasonable measurement accuracy. There is a slight anomaly for the larger 2 Oz traces at the shortest times caused by measurement uncertainties in that region. The current levels are very high and the sensitivity to current level very low in that region for those sizes.

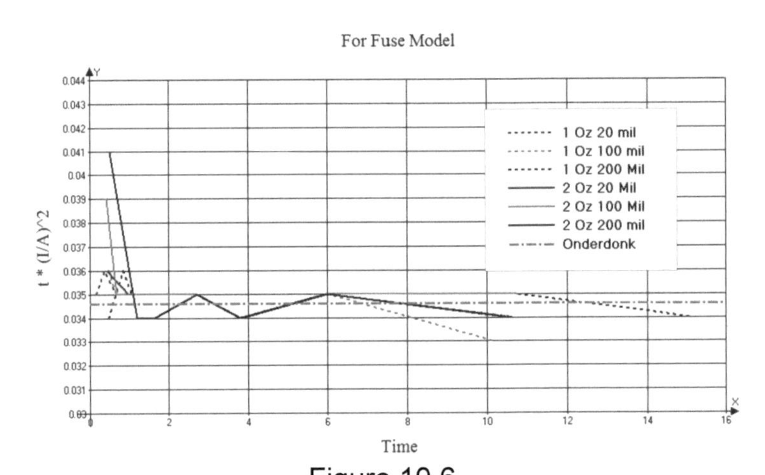

Figure 10.6
Plot of $t^*(I/A)^2$ for the various configuration simulations of the fuse trace.

When we plot the same term for the TRM-Trace model simulations, we get the interesting result shown in Figure 10.7. They are all upward sloping to the right, a clear indication of the cooling effects that are taking place with time.

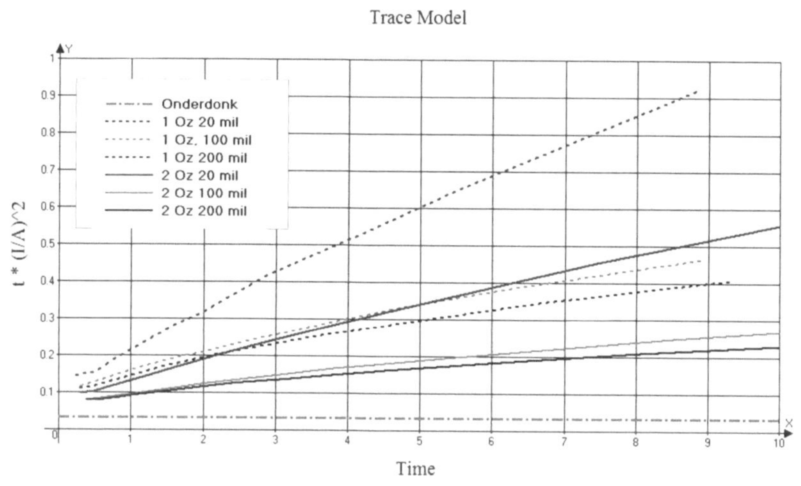

Figure 10.7
Plot of $t^*(I/A)^2$ for the various configuration simulations of the trace model,.

But when we look closely at these same curves at very short times, they all seem to originate at a similar value (Figure 10.8.)This is not too far different from our fuse simulation. Therefore, at the first instant of time, all the trace configurations tend to start out the same. Furthermore, except for the smallest trace (1 Oz. 20 mil), they are all at or under 0.25 during the first three seconds.

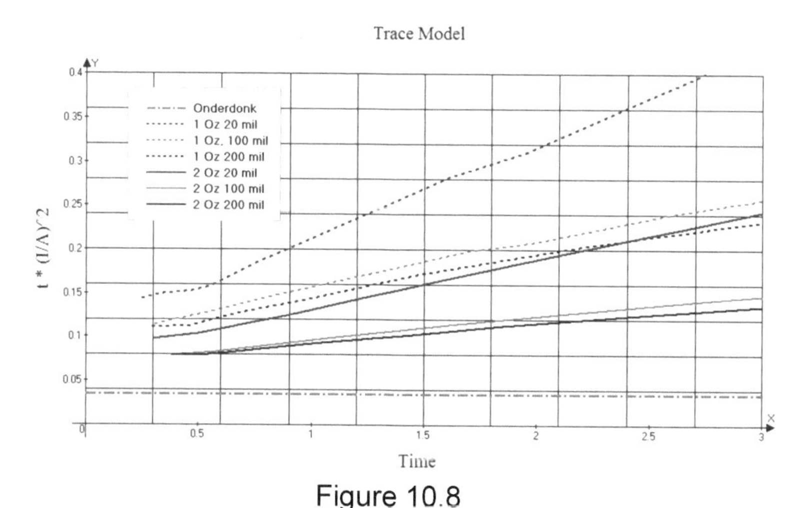

Figure 10.8
Plot of $t^*(I/A)^2$ for the various configuration simulations of the regular traces at times under three seconds.

It is instructive to compare the fusing current for the traces at various times to the fusing current calculated from Onderdonk's equation. I calculate the ratio of these two values and provide them in Table 10.1.

Ratio of Trace Model Current to Onderdonk Current								
			time (Seconds)					
Trace Config		Area	0.5	1	2	3	4	5
Oz.	Width							
2	20	52	3.46	3.87	4.65	5.3	5.84	6.3
1	20	26	2.24	2.45	3.03	3.53	3.83	4.2
1	100	130	2.02	2.14	2.45	2.69	3.02	3.15
1	200	260	1.88	2.04	2.37	2.62	2.77	2.97
2	200	520	1.57	1.67	1.9	2.05	2.2	2.37
2	100	260	1.51	1.63	1.82	1.98	2.1	2.2

Table 10.1

A very interesting result here is that, relatively speaking, narrower traces take longer to reach the fusing temperature than do wider traces. That seems, on the surface, to be counter-intuitive. But the narrower traces do cool better. Figure 10.9 gives us a clue as to why. The top thermal graph is from underneath the 1 Oz. 20 mil trace when it reaches the fusing temperature. The lower thermal graph is the same thing for a 2 Oz. 200 mil trace. The cooling "plume" is the dark blue area around which the trace is "shedding" heat. Comparing the area of this "plume" to the trace width illustrates that the narrower trace has, relatively speaking, a much wider plume *compared to its width* than does the wider trace. The temperature is therefore cooler under the narrower trace (720° C compared to 952° C.)

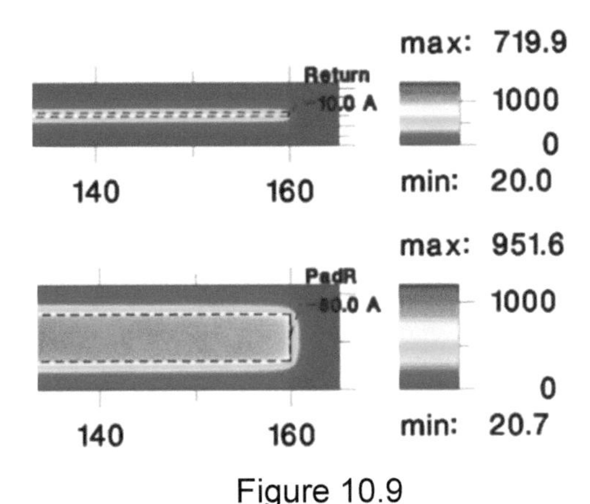

Figure 10.9
Comparing the thermal "plume" under two traces of differing width.

Referring back to the question first posed in this paper (how large a trace is needed to carry 40 Amps for 1 second), we can approximate an answer to that question from Figure 10.8 and Table 10.1.

We assume the trace will be at least 100 mils wide
Calculate t $(I/A)^2 < .15$, or $(1)*(40/A)^2 < 0.15$ or $(40/A)^2 < 0.15$.
This leads to: $A > (1600/0.15)^{.5} = 103$ square mils
A 1Oz 76 mil wide trace has an approximate area of 103 sq. mils.

We can check the answer by calculating the fusing current for this result using Onderdonk's equation and then looking at Table 1.

Onderdonk: $t = .0346*(A/I)^2$ (From [Eq. 10.3])
This leads to $I^2 = .0346(A^2)/t = .0346(103)^2$
Leads to I = 19 Amps
From Table 1, a 130 sq mil area trace model will have a 2.14 times greater current carrying capacity, so the current carrying capacity of the trace would be about 2.14*19 = 40.6, roughly what we wanted.

10.6 Final Conclusions

Looking at the question of fusing currents on PCB traces can be very problematic. Onderdonk's equation is a reasonable place to start, and thermal simulation models suggest that the actual fusing times can be 1.5 to 6.0 times longer than Onderdonk suggests (and more considering the restrictive assumptions of the models.) But there are many, subtle variables involved. It is not really practical to solve this type of problem with a formula or graph. The use of a simulation model is almost required; and even more so if there are adjacent traces or planes that can contribute to trace cooling.

And a word of caution. If a trace fuses, it is *never* acceptable to repair the trace and put the board back into service. If a trace fuses, that is considered a destructive failure and the board must be removed from service. The heat from the fusing temperature may have caused any number of hidden potential problems in the material and components around that area. It may be useful to use a trace as a fuse, but doing so is a one-time event. If a replaceable fuse is required, a fusing component must be used instead.

Notes:

1. A circular mil is the area of a circle with a diameter of one mil. The formula is $A = d^2$. The conversion from circular mils to mil^2 is $\pi/4$. Normal conversions are:

   $1\ mil^2 = 1.273$ circular mils

   1 circular mil = $.7854\ mil^2$

   $1\ m^2 = 19.736*10^8$ circular mil

   1 circular mil = $5.067*10^{-10}\ m^2 = 5.067*10^{-4}\ mm^2$

2. From a thermodynamic engineering standpoint, there is a significant difference between the ambient temperature and the initial condition. For example, there could be some current-induced heating of a wire above the ambient temperature in a normal operation before a change of current is applied. In the electronic industry we are less rigorous and tend to refer to the initial temperature of the trace as the "ambient" temperature. In this chapter we are assuming that the initial condition (temperature) is the same as the ambient temperature. If the initial temperature of the trace were different from the temperature around the trace, we would refer to the initial trace temperature as the "ambient" temperature.

3. Furthermore, if we let A = constant, then Equation 10.4 reduces to I^2t = constant. The I^2t value is an important rating in the fuse industry. See http://en.wikipedia.org/wiki/Fuse_%28electrical%29#The_I2t_value

Supplemental Comment:

Recently a person contacted me with a different kind of requirement. It was:

I need my trace to:

1. *Carry 2A on turn-on, but only for 2 to 4 seconds.*
2. *If there is a dead short, carry 60 Amps for 0.150 sec (150 ms).*
3. *Worst case is a stalled motor, 8-20 A for 20-30 seconds.*
4. *Trace is on an internal layer.*

How big a trace do I need?

These kinds of requirements, while not too common, are more common than you might think. I did not have the resources to help work with this person to find an answer.

11 FUSING CURRENTS: EXPERIMENTAL VERIFICATION

11.1 Problems in Predicting Fusing

In Chapter 10 we used the TRM thermal simulation program to simulate the fusing characteristics of a fuse and of a typical PCB trace. The fuse was pretty straightforward, although pretty theoretical. But when it comes to simulating an actual trace, there are some significant problems we need to be aware of. Almost all of these potential problems are caused by the fact that we are covering a *very* wide temperature range, from around 20° C to 1,083° C. Some of these problems are covered below.

11.1.1 Heating: Traces heat by I^2R power dissipation in the trace. Power is also given by v*i (voltage times current.) Under experimental conditions, both voltage and current can be measured with pretty high precision, so this is typically not too big a problem. The resistance of the trace can be found by dividing voltage by current (v/i). The change in temperature is given by the same formula we used in Chapter 4 (Equation 4.4), repeated here as Equation 11.1.

$$\Delta T = \frac{1}{\alpha_0}\left[\frac{R_t}{R_{t_0}} - 1\right]$$

[Eq. 11.1]

Where:

- ΔT = change in temperature from ambient (ref) (°C)
- α = thermal coefficient of resistivity of copper (1/°C)
- Rt = resistance at the temperature under test (Ohms)
- R_{to} = resistance at the reference temperature (Ohms)

It should be noted that this type of analysis leads to the *average* temperature of the trace, which, because of thermal gradients (especially at the ends of the trace), is not the maximum temperature anywhere along the trace.

Computer models typically assume that α is either a constant or that it changes linearly with temperature, and also that resistance changes linearly with α. As traces heat to relatively high temperatures, neither of these assumptions may continue to be true.

11.1.2 Cooling: A PCB trace typically cools primarily by conduction through the dielectric and secondarily by convection and radiation into the air. A stable temperature is reached when heating equals cooling. Computer models are reasonably good at handling the conductive cooling as long as the thermal properties of the material are known and predictable (I.e. as long as the material properties don't change.) If the thermal properties are not predictable, then the computer model can no longer provide reliable results.

Convection and radiation are typically handled in thermal simulation models by a factor called HTC, Heat Transfer Coefficient. The values of HTC are reasonably well understood for temperature changes in the range of up to 100 degrees C or so. But they are very difficult to predict if there are hot spots or for a temperature range of 1,000 degrees C, especially if the timing associated with that range of temperature is uncertain.

11.1.3 Trace (copper) properties: There are two types of copper we typically have on PCBs. There is the copper foil that is bonded to the dielectric material in the supply process, and there is copper plating applied by the board fabricator (typically on the outer layers.) Since "copper is copper," it is tempting to think that a model could consider copper traces to be homogeneous. But let's take a closer look:

Copper foil is fabricated under petty well-controlled conditions. There might be a tolerance on the foil thickness, but otherwise, it should be pretty homogeneous. On the other hand, there might be microscopic contamination that exists at places under the foil that happens during the bonding process. If these change the heat transfer characteristics between the foil and the dielectric, then unpredictable thermal hot-spots might develop at these points.

Copper plating is a process that is inherently less well controlled. Plating can (and does) vary in thickness at places on the board (and therefore possibly at places along a trace) and there are more opportunities for contamination to occur. Again, these can lead to hot-spots along a trace.

Finally, final trace dimensions are defined by a photo-etch process. There can be small variations in trace width along a trace. Narrower places along a trace would heat more quickly.

11.1.4 Dielectric Properties: Dielectric materials, at least from reliable manufacturers, are pretty homogeneous. However, typical materials (such as FR4) have three parameters of interest here. They were covered in Section 2.2, above, and will be repeated briefly here.

Tg: Glass transition temperature is the temperature region where the polymer transitions from a hard, glassy material to a soft, rubbery material. This results in dimensional changes and some change in thermal conductivity. This changes the way in which the dielectric is able to conduct heat away from the trace. It is reversible. A typical specification for Tg is around 170 degrees C.

Td: The thermal decomposition temperature is the temperature at which weight loss begins to occur. Thermal decomposition is typically specified around 300 to 350 degrees C. Thermal decomposition will change the ability of the dielectric to carry heat away from the trace. It is irreversible.

Delamination: In addition, if the board begins to delaminate, all bets are off, and the trace will begin to heat uncontrollably and probably fairly rapidly. The material used in the evaluation to follow had a "Thermal Resistance" (time to delamination) specified as:

at T = 260 oC   > 60 minutes
at T = 288 oC   > 20 minutes

11.2 The Fusing Process

The fusing process can be subdivided into two broadly defined categories, strong overload (leading to rapid fusing), and long-term, or slight, overload (leading to slow fusing that might take minutes or hours.)

11.2.1 Strong Overload: If we submit a trace to a strong overload (relatively large current), the temperature rises quickly to the point where the weakest spot on the trace melts. The weakest spot on the trace is unpredictable and will typically vary from trace to trace. The melting process will be rapid, particularly because there is "positive" feedback. That is, the rising current results in rising temperature, which raises the resistance, which increases the temperature, which raises the resistance....

11.2.2 Slight Overload: If we apply just enough current to ultimately cause the trace to melt, the failure process goes through several separate steps. First the trace temperature will rise to a plateau and seem to level off. If this is near Tg, the temperature will slowly increase until it gets in the vicinity of the delamination temperature. The temperature will start increasing a little faster until it gets near Td. At that point the temperature really accelerates until the trace melts at the weakest spot along the trace. The entire process can take from tens of seconds to hours.

If we plot trace temperature against time, we can't necessarily identify the precise times at which these transitions occur. They will occur at points within, or perhaps underneath, the copper trace itself and by their very nature are "point" phenomena (remember, the calculated trace temperature is an *average* over the entire trace length, not the temperature at the hottest point.)

11.3 Experimental Results

As a result of Prototron's generosity (see Acknowledgement), there were several boards available for evaluation. The test equipment available included a constant current power supply and a portable oscilloscope. The maximum current capacity of the power supply was 10 Amps, which limited the investigation to smaller traces. Separate evaluations suggested the accuracy of the power supply and of the scope were within 2 or 3 percent.

Traces were 6 inches long and varied in trace width from 15 mils to 27 mils. Most were 0.5 Oz foil plated over with 1.0 Oz copper, although there was one board that was 0.5 Oz foil only. Separate microsection measurements showed that trace widths were well controlled, but the total foil plus plating thickness could vary between 1.9 mils and 2.6 mils, worst case. In this chapter I will report on three cases.

11.3.1 Case A (Fast Fusing): A 15 mil wide, 0.5 Oz trace was subjected to a sudden current of 6 Amps. The trace fused in 2.75 seconds (see Figure 11.1.) The curve of voltage vs. time ramps in an almost straight line. Temperature lines are shown on the curve for reference. They are approximate [1].

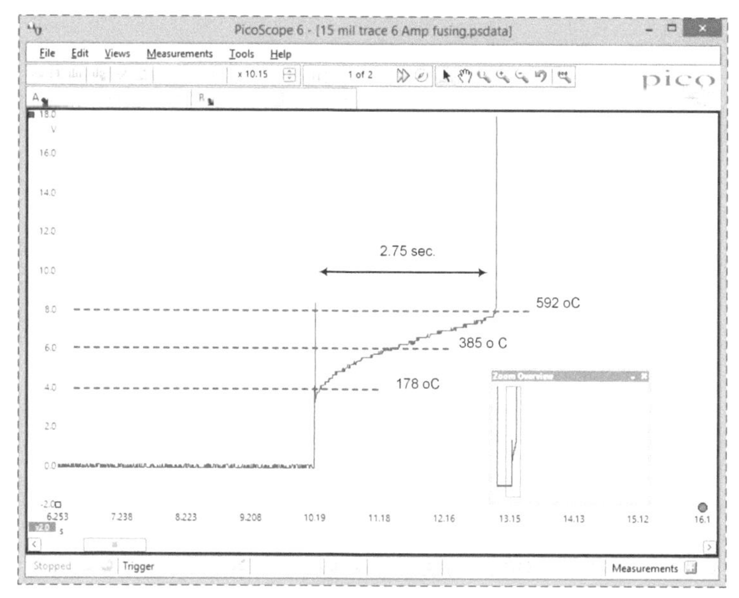

Figure 11.1
15 Mil wide, 0.5 Oz trace carrying 6 Amps.

A video was taken of the trace fusing. The moment of fusing was really somewhat underwhelming! Fusing occurred within one video frame (1/30 th second, see Figure 11.2)! There was no smoke and virtually no aroma. There was negligible damage to the board under and around the trace.

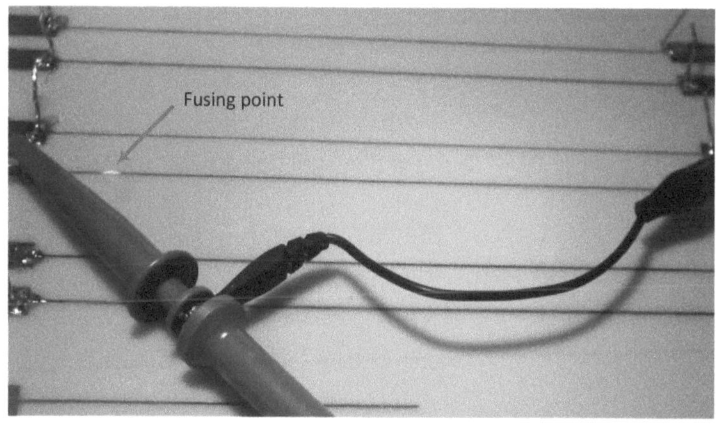

Figure 11.2
15 mil wide, 0.5 Oz. trace at the moment of fusing.

A thermal simulation of this trace calculated the fusing time at 4 seconds. This is pretty close and is likely the fusing time had the trace and board been homogeneous. Results are partially predictable because the heating happens so quickly there is little opportunity for any cooling discontinuities to develop. This is evidenced by the lack of any visible damage to the board material. But as discussed above, there are inherent potential hot spots related to the copper that can cause localized heating. The trace fuses at its weakest point, and that point is much hotter than the surrounding trace. There is no way to predict or explain why that particular point was the weakest point. The average temperature of the trace is still pretty low when the trace fuses at its weakest point.

11.3.2 Case B (Slow Fusing): A 20 mil wide, 1.5 Oz trace was subjected to 8.5 Amps. This trace, being 33% wider and 3 times thicker than the case above, took much longer, about 30 minutes, to fuse. The curve of voltage (and therefore temperature) vs. time is shown in Figure 11.3.

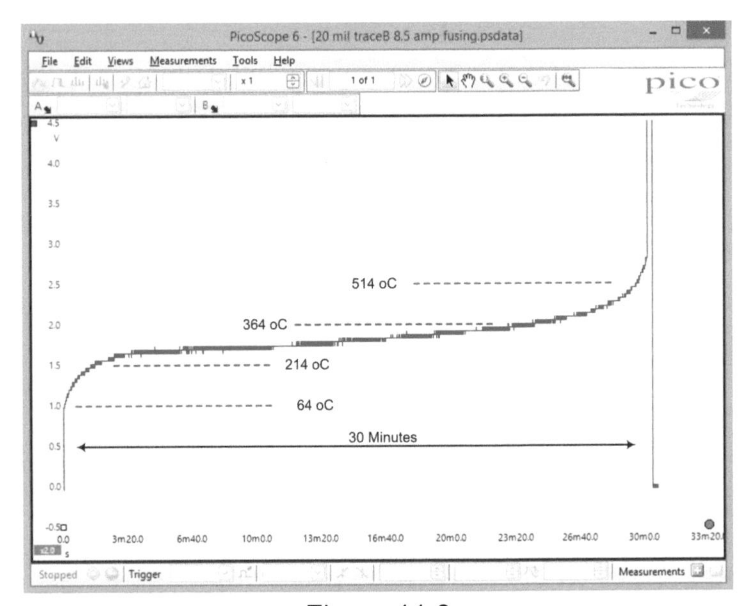

Figure 11.3
20 mil wide, 1.5 Oz. trace carrying 8.5 Amps.

Here the curve tends to level out somewhere around maybe 250 degrees or so and then starts a slow climb upwards. As the temperature increases the rate of increase also increases. Some people have described this as a thermal runaway. Once it starts its progress is inevitable (unless the current is interrupted.)

When this trace fused the results were much more spectacular. Figure 11.4 illustrates the moment of fusing. First, the trace began to visibly glow about 35 seconds before fusing. It began smoking some 90 seconds before fusing. The aroma of burning material was in the air for perhaps 15 minutes before fusing.

Note in Figure 11.4 how smoke is being visibly ejected under pressure from several places along and under the trace. The board shows visible signs of heat damage, and closer inspection shows continuous areas of what appear to be bubbles (delamination) along the edges of the trace.

Figure 11.4
Moment of fusing for a 20 mil wide, 1.5 Oz trace
carrying 8.5 Amps.

11.3.3 Other Cases: The current required for the kind of "thermal runaway" described in Case B, above, is very trace specific. This is because the discontinuities associated with each trace and each spot on the board are different. For example Figure 11.5 and Figure 11.6 illustrate the same trace carrying 8.3 Amps and 8.4 Amps respectively. (The 8.4 Amps was applied 12 hours after the 8.3 Amps, allowing plenty of time for cooling between tests.) At 8.3 Amps, the trace showed no signs of running away after 2 hours. The same trace carrying 8.4 Amps fused at 1 hour and 16 minutes. Another trace on another board with the same nominal dimensions and carrying 8.5 Amps showed no signs of runaway for over an hour, illustrating how there can be minor but important differences between traces and boards.

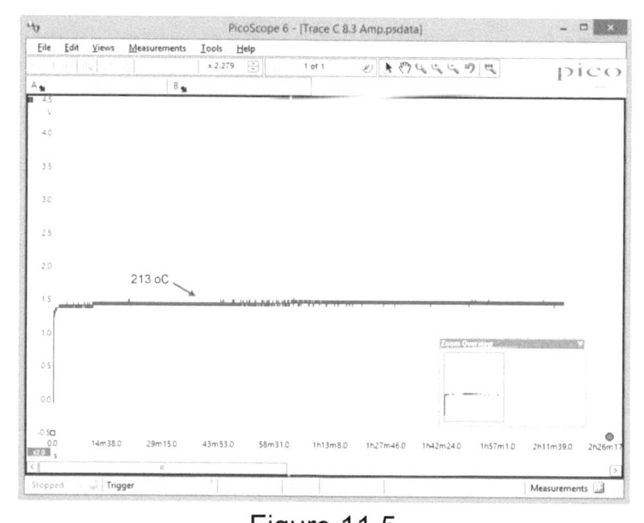

Figure 11.5
20 mil 1.5 Oz trace carrying 8.3 Amps showed
no signs of runaway after 2 Hours.

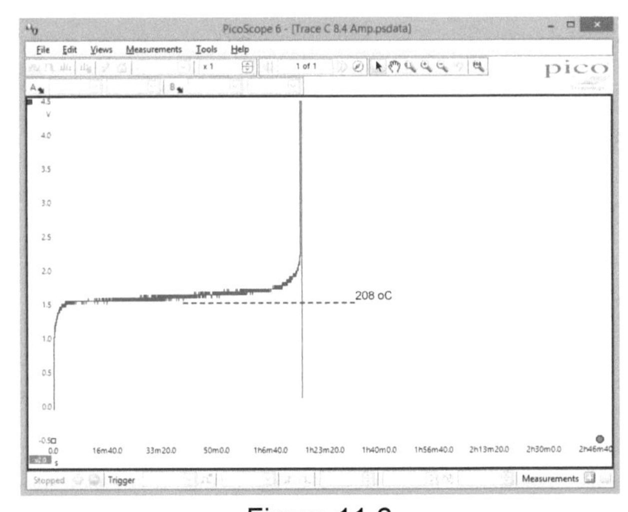

Figure 11.6
Same trace carrying 8.4 Amps fused in 1 hour, 16 minutes.

11.4 Summary

The fusing time of a trace subjected to a heavy current overload (and fusing between 0.0 and maybe 5 seconds) is modestly predictable. Under these conditions there is negligible time for any cooling effects to kick in and the board material properties remain constant. The trace fuses very quickly at a specific point and there is little damage to the surrounding board area. There is little aroma associated with type of failure.

Slight current overloads result in unpredictably long fusing times (as long as hours). This type of overload might occur, for example, if a trace relies on supplemental cooling processes --- say flowing air or an embedded heat sink --- and that process degrades for any reason. The board material shows significant damage all along the trace, and smoke is apparent. Some smoke is ejected from under the trace under significant pressure. The trace may glow red hot for several seconds before it fuses. There is heavy material damage under and around the trace. When the trace finally fuses, it does so along a small distance (not at a specific point), the copper seeming to "curl back" as it melts. The aroma of burning is pervasive.

Nominally identical traces have different currents where the slight overload (sometimes referred to as thermal runaway) will result in a fusing condition. This is because of unpredictable discontinuities in the board material and the copper composition.

Notes:

1. Temperature reference lines in Figures 11.1, 11.3, 11.5, and 11.6 are estimated. The reference resistance (at 20° C) was calculated using nominal trace geometry and a resistivity of 0.67 uOhm-in. Resistance at temperature was calculated from v/i (Ohms Law). The elevated temperature was then calculated using Equation 11.1 using the thermal coefficient of resistivity as 0.0039. Remember, these are *average* trace temperatures along the trace, not the peak temperature at any point.

12 DO TRACES HEAT UNIFORMLY?

As discussed in Chapter 11, and shown in Figures 11.2 and 11.4, when a trace is heated near the fusing point hot spots develop along the trace. The traces tend to fuse (melt) at a *point* along the trace, not along a broad area.

There is an interesting figure in a 2010 paper [1] on fusing (Figure 12.1) that shows hot spots developing on traces. The caption for this figure states *"...PCB trace over 300°C and starting to have hot spots, which represent possible fusing locations."*

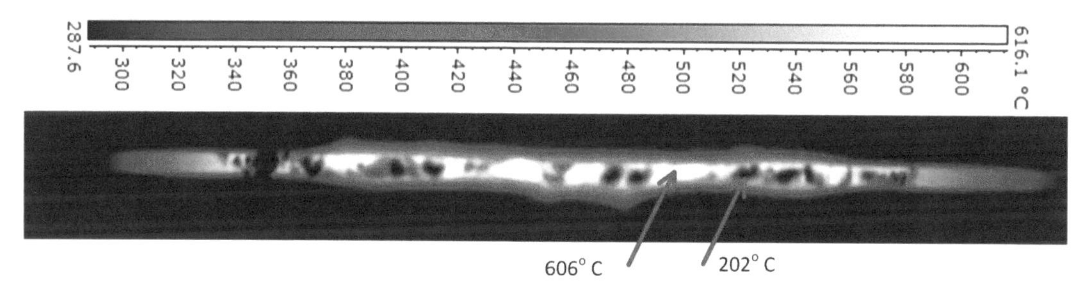

Figure 12.1
Hot spots on a 1.0 Oz., 80 mil wide trace just before fusing.

I saw a similar pattern in my own fusing investigations (Chapter 11). It is shown in Figure 12.2.

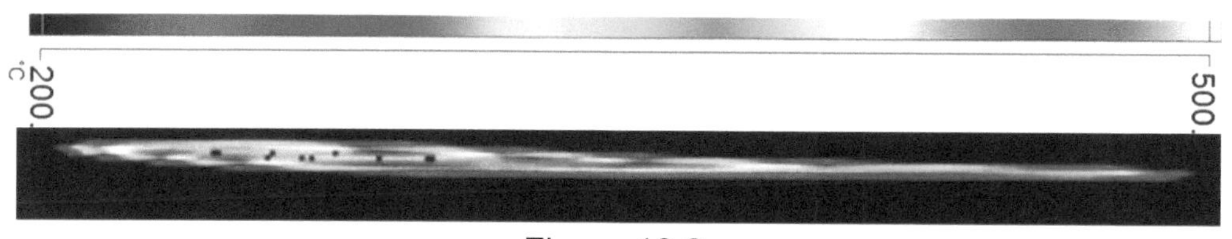

Figure 12.2
20 mil, 0.5 Oz. trace heated to a maximum of 436° C.

In my own testing of trace temperatures using a thermocouple, I noticed that if I moved the thermocouple slightly (right or left) the indicated temperature might change by as much as 1.5 degrees. This is more than the measurement tolerance (I believed) but still not enough to be able to say whether the observed result was real, or still just an indication of measurement tolerance.

We would expect copper traces to heat relatively smoothly along their length. That is, any temperature gradients would be expected to be smooth. Clearly this expectation breaks down when the trace is heated hot enough. But all the above raises the question: When do "hot spots" begin to develop along a trace? In particular, do traces, even at low temperatures, exhibit uneven temperatures along their length?

12.1 Thermal Gradients on Traces

In order to answer this question, I needed access to a thermal imager. Scott Dau, a Seattle Firefighter and part-time fire investigation instructor, loaned me a Fluke model Ti32 thermal imager. Thermal imagers work on an infrared sensing technology, and measure temperatures by measuring the target's infrared characteristics.

In order to achieve accurate measurements, the target's *emissivity* [2] must be known, or at least controlled. One way to do this is to paint the target (in this case, the traces and surrounding board area) with a flat black paint. The relevant portions of the test boards were painted a flat black with a spray enamel.

Figure 12.3 illustrates the thermal profile of a 100 mil wide, 0.5 Oz foil (not plated) trace heated to a maximum temperature of 166° C. It appears about as expected. But note that the hottest spot is not at the center of the trace, but nearer the left side. This is consistent with where the traces in Chapter 11 fused.

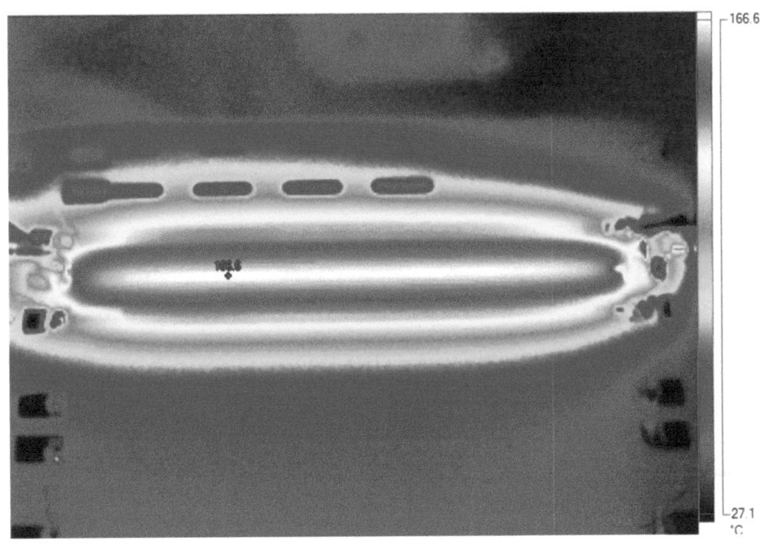

Figure 12.3
Thermal image of 100 mil wide, 0.5 Oz. trace, heated to about 166 degrees C.

But what is really interesting is if we narrow the thermal range of the image to 150.0 to 166.6 degrees C (Figure 12.4). This has the effect of giving us a thermal "close-up" of the trace. Now it becomes apparent that the heating of the trace is not at all uniform along its length.

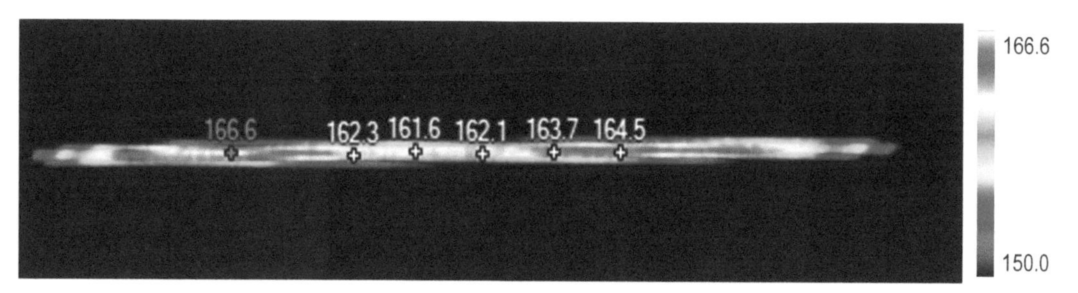

Figure 12.4
Thermal close up of Figure 12.3.

But what if the temperature of the trace is raised to a more modest level? Figure 12.5 shows the thermal distribution of the same trace with the maximum temperature raised to only 49.6° C. The non-uniformity of the temperature distribution is evident even over this lower thermal range.

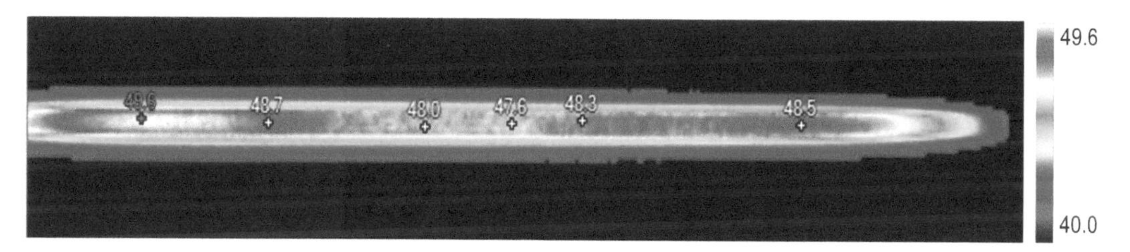

Figure 12.5
Thermal close up of Figure 12.3 raised to a maximum temperature of only 49.6 degrees.

Note also in Figure 12.5 that the trace is cooler along the outside edge than it is down the centerline. This is true in virtually every image. The reason is that the cooling path from the center of the trace (out to the edge) is longer than it is from the edge of the trace.

12.2 Thermal Gradients on Narrow Traces

In the above section we looked at 100 mil wide traces. What about smaller traces? Figure 12.6 illustrates the thermal profile of a 20 mil wide, 0.5 Oz. trace, raised to a maximum temperature of 62.4° C. It exhibits a non-uniformity very similar to the wider, 100 mil trace.

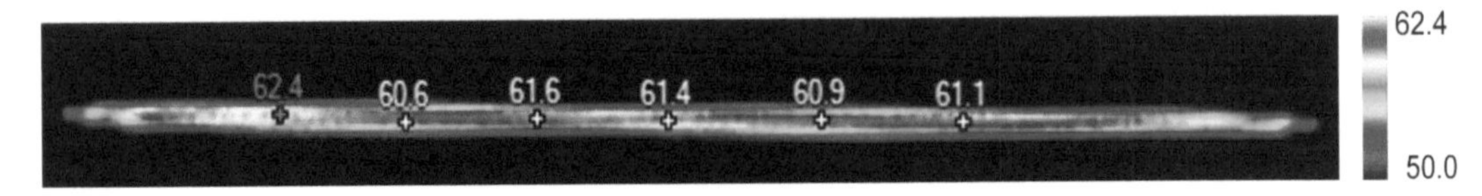

Figure 12.6
Thermal profile of a 20 mil wide, 0.5 Oz. trace raised
to a maximum temperature of 62.4 degrees.

12.3 Parallel Investigation

While researching this topic, I entered into a collaboration with Norocel Codreanu, another researcher in Romania [3]. He was interested in this same topic and ran a series of tests on some traces there. One of his results is shown in Figure 12.7 for a 100 mil, 1.0 Oz. trace carrying 5 Amps of current.

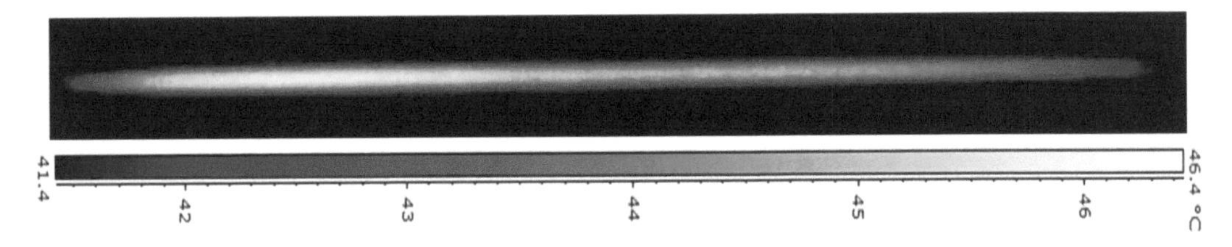

Figure 12.7
Independent investigation of 100 mil wide, 1.0 Oz. trace.

His investigation did not show the same non-uniformity of heating, perhaps because of the different manufacturing method he used.

12.4 Does Trace Thickness Matter?

As a result of other investigations, I had a variety of trace thicknesses I could evaluate. They ranged from a simple, 0.5 Oz, 100 mil wide trace (nominally 0.65 mils thick) to a complex 2.0 Oz., 200 mil wide trace plated up with an additional 2 ounces of plating, for a nominal total of 5.2 mils thickness. The various combinations are shown in Table 12.1.

The results from these seven options are shown in Appendix A9. They all show varying degrees of heating non-uniformity, but I was not able to discern any particular patterns related to trace thickness or degree of copper plating on top of the foil.

Trace options available for thermal hot-spot analysis					
Width mil	Foil (Oz.)	Plating (Oz.)	Total (nominal) thickness (mil)	Current thru trace (A)	Max Temp (degrees C)
100	0.5		0.65	5.6	53
100	0.5	1.0	1.9	8.0	54
100	1.5		2.0	7.5	54.5
100	1.5	1.75	4.4	10.0	47
200	2.0		2.6	10.0	39
200	2.0	1.0	3.9	10.0	35
200	2.0	2.0	5.2	10.0	33.6

Table 12.1

12.5 What Causes Thermal Non-uniformity?

At this point it is unknown exactly what causes the thermal non-uniformity in the heating of the traces. There are a variety of possibilities:

1. Contamination under the foil or under the copper plating (if any)
2. Contaminants in the copper alloy itself.
3. Variations in trace thickness or width due to normal manufacturing tolerances.
4. Variations in the properties of the underlying dielectric material.

12.6 Conclusion

It is clear that traces do not heat uniformly once high levels of temperature are reached. The fact that traces at high temperature fail at random points is clear evidence of this.

It also seems apparent that traces do not heat uniformly even at lower temperatures. It can be stated with some confidence that traces are hotter down the centerline and they cool towards the edges, primarily because the cooling path is shorter the nearer one is to the edge.

We can say with less confidence that traces at lower temperatures do not heat uniformly along the center-line. Whether this happens seems to depend on many not-yet-understood parameters, particularly since Norocel Codreanu's investigations failed to detect much non-uniformity. There is still more research to be done in this area.

Notes:

1. "New Methods of Testing PCB Traces Capacity and Fusing", Norocel Codreanu, Radu Bunea, Paul Svasta, "Politehnica" University of Bucharest, Center for Technological Electronics and Interconnection Techniques, UPB-CETTI, internal research project and report of UPB-CETTI / Winter-2010, link: http://www.cetti.ro/cadence/articles/New_Methods_of_Testing_PCB_Traces_Capacity_and_Fusing.pdf
2. See https://en.wikipedia.org/wiki/Emissivity
3. See Norocel's affiliation in Note [1]. Norocel used traces fabricated in his own lab using a standard sheet of plated dielectric and the "Press-and-Peel" blue foil process. The thermal images were taken with a Flir ThermaCAM model SC640.

13 AC Currents

In previous chapters, all models and experimental results were based on switched, constant currents. In this chapter we will look at what happens when we apply various types of ac currents to a trace.

13.1 Basic Models:

One, relatively simple, way to utilize TRM to evaluate ac current is to construct a basic model [1] of a trace and apply the appropriate current to it. The trace model we have chosen is the same trace we used in our experimental via analysis (Section 7.3), a 27 mil wide, 2.9 mil thick trace on a dielectric with known thermal conductivity (0.7 W/m-K), mass density (2000 kg/m³) and specific heat capacity (900 J/kg-K). We will apply 5.0 Amps DC to the trace as a reference. We will then assume the applied current is an AC square pulse with a varying duty cycle. The appropriate equivalent DC current for each duty cycle is the effective RMS current (for a square wave), given by the formula RMS = 5 * SQRT(DCy) where DCy is the duty cycle expressed as a fraction. Table 1.31 provides the detailed equivalent currents and the maximum temperatures from the TRM model for that equivalent RMS current. (These data calculations include the effects of the thermal coefficient of resistivity. [2])

Duty Cycle (%)	RMS Current (A)	Max. Temp °C
100	5.000	65.7
99	4.975	65.3
90	4.743	61.2
80	4.472	56.8
70	4.183	52.6
60	3.873	49.5
50	3.536	45.4
40	3.162	41.2
30	2.739	37.3
20	2.236	33.8
10	1.581	29.8
1	0.050	26.0

Table 13.1
Equivalent RMS currents and TRM temperature calculation.

The result is a linear curve as shown in Figure 13.1 [3]. The maximum trace temperature is 65.7 degrees C, with 100% duty cycle (i.e. direct current.) It declines in a linear fashion to the ambient room temperature at a duty cycle of zero (i.e. no current.) The ambient temperature is set to 26 degrees in this model for reasons that will be clearer below.

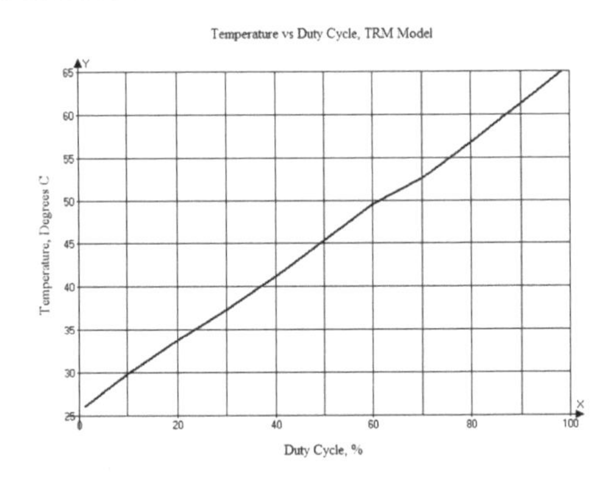

Figure 13.1
Trace temperature in our model from a 5.0 Amp pulse as a function of duty cycle.

Figure 13.2 illustrates a graph of temperature vs time for the 5.0 Amp DC current (black line) and for 3.536 Amp DC (dotted red line), which is the RMS equivalent current for the 5.0 Amp current at a 50% duty cycle. These curves will rise asymptotically to the expected values of 65.7 degrees C and 45.4 degrees C, respectively. The scale on the right hand axis is the equivalent RMS current (Amps) that will lead in the long term to the corresponding temperature on the left axis.

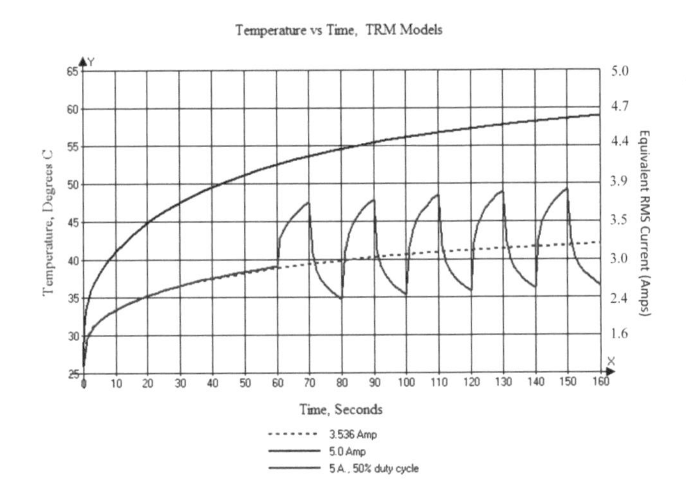

Figure 13.2

Temperature curves for a 5.0 Amp, 0.05 Hz waveform at 50% duty cycle (red) along with 5.0 Amp (black) and 3.536 Amp (dotted red) curves.

The solid red curve in Figure 13.2 is the resulting modeled waveform at 50% duty cycle. Its "mean" temperature is the same as that found in Figure 13.1, trending to about 45.4 degrees C.

TRM offers an alternative way to look at an alternating waveform. That is to apply current to the waveform according to a cyclical schedule supplied by the user. The schedule is provided by way of an Excel spreadsheet. An example is shown in Table 13.2. Time is the time in seconds. "Power" and "Return" are the current input and output pads, respectively, for the model. The values in the table are the current that will be applied to those pads until the next time specified. This example is for a 5.0 Amp current (RMS value is applied for the first 60 seconds [4]) at a 0.05 HZ frequency (after the first 60 seconds) with a 50% duty cycle. The output result is plotted as the solid red curve in Figure 13.2.

time	Power	Return
0	3.5355	-3.5355
60	5	-5
70	0	0
80	5	-5
90	0	0
---	---	---
150	0	0
160	5	-5

Table 13.2
Cyclical input data for TRM analysis

TRM offers yet a third way to look at all this. First we run a transient model at 5.0 Amps (the black curve in Figure 13.2) which derives a "heating curve." Then we use the PWM (Pulse Width Modulation) option found under the "Extra" menu (Figure 13.3). This modulates the heating curve with a duty cycle. Setting the "Time interval t2" and "Time interval t3" each to 10 ten seconds gives us a 20 second cycle or 0.05 Hz. The 1.464 Watt power value comes from a basic model run of 5.0 Amps. It is the power dissipated in the trace (= I^2*R). [5]

Figure 13.4 is the resulting waveform. The shape of this waveform looks almost exactly like the red waveform in Figure 13.2. (Technical note: The PWM approach will slightly understate the other approaches in amplitude because it does not yet take the thermal coefficient of resistivity into consideration for the calculations. See again [2].) Therefore, these latter two approaches to determining the waveform of an AC signal are equivalent, but the last approach is the easiest and most versatile, and requires minimal CPU time.

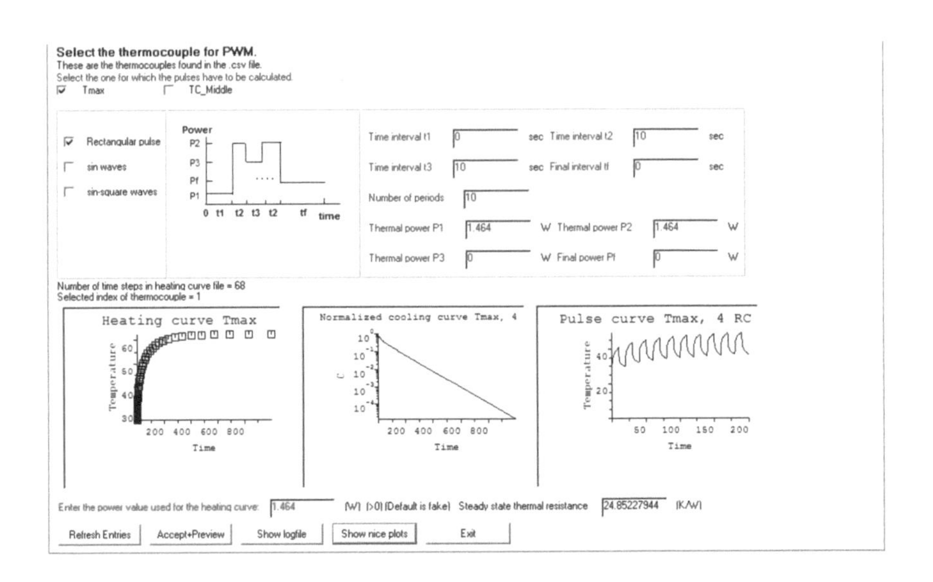

Figure 13.3
Setting up the PWM function in TRM

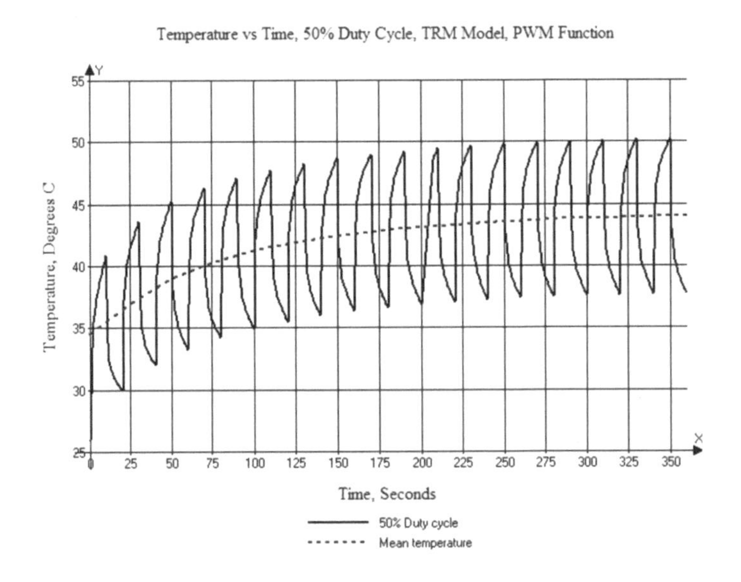

Figure 13.4
The resulting waveform calculation from the PWM function

This now gives us an easy way to look at other duty cycles. For example, let's set t2 and t3 to 5 and 20 seconds, respectively (and vice versa). This gives us a frequency of 0.04 Hz, at a 20% or 80% duty cycle. This results in the curves shown in Figure 13.5. TRM's PWD function also provides the mean temperature for each cycle in this waveform. The trend line for the mean values approaches the 33.8 degrees C maximum temperature for the 20% duty cycle and the 56.8 degree C maximum temperature for the 80% duty cycle provided in Table 13.1 (see again [2]).

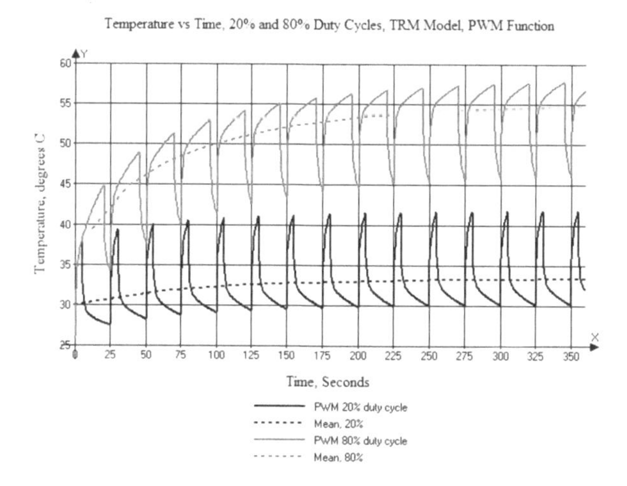

Figure 13.5
The resulting waveform calculation from PWM for a 0.04 Hz frequency, 20% and 80% duty cycle waveforms.

So far, the TRM models seem to make some intuitive sense at lower frequencies. The PWM function will also enable us to estimate the temperatures of traces at higher frequencies. Figure 13.6 is the result from the PWM function at 100 Hz, 50% duty cycle. The cycles are now so short that the individual cycles are not visible, and have no effect on the result. The curve looks like the curve for a DC current at the equivalent RMS level (3.54 Amps) and levels off at about 44 degrees C. This equates to a value from Figure 13.1 of 45.4 degrees (but without the thermal coefficient of resistivity adjustment.)

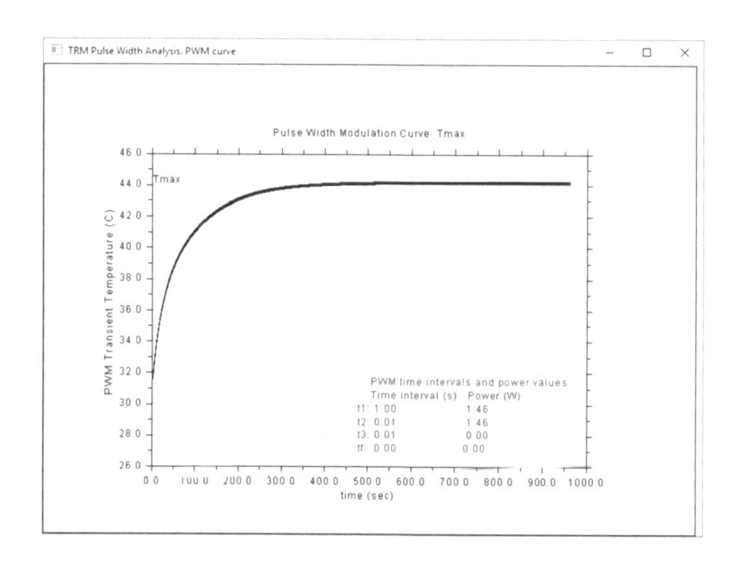

Figure 13.6
PWM result for a 100 Hz, 50% duty cycle input current.

13.1.1 Preliminary results: The conclusions so far would seem to suggest:

1. The temperature of a trace is determined by the equivalent RMS value of the current driving it.
2. The temperature of a trace at any duty cycle seems to be independent of frequency (at least at frequencies below where the skin effect comes into play.)

13.2 Experimental Verification:

We had two boards left over from our via evaluation (see again Section 7.3) to use in our experimental testing. One of the traces on each board is identical to the model used above. The diagram of the test procedure is shown in Figure 13.7.

A constant current generator is set to 5 Amps. It is important that the current flow not be interrupted, otherwise the generator output will change dramatically, trying to maintain the constant current flow. Therefore, it is important to maintain a constant current from the generator, even though the current through a trace is changing. To achieve this, we used two identical test traces with current switching between them very quickly. The current was switched by a pair of Darlington transistors, each driven by a waveform generator. The waveform generator has two complementary outputs. These outputs were set to a square wave with output between 0.0 and 4.0 volts. The frequency was adjustable between 0.01 Hz and over 20.0 MHz in 0.01 Hz steps. The duty cycle could be set from 0.1 % to 99.9% in 0.1% steps. Current and voltage meters, and oscilloscope probes, were set where indicated. The temperature of each trace was measured with a thermocouple data logger with a sampling rate of 1.0 second and a resolution of 0.5 degrees C. The ambient environment was a typical office environment with temperature about 26 degrees C.

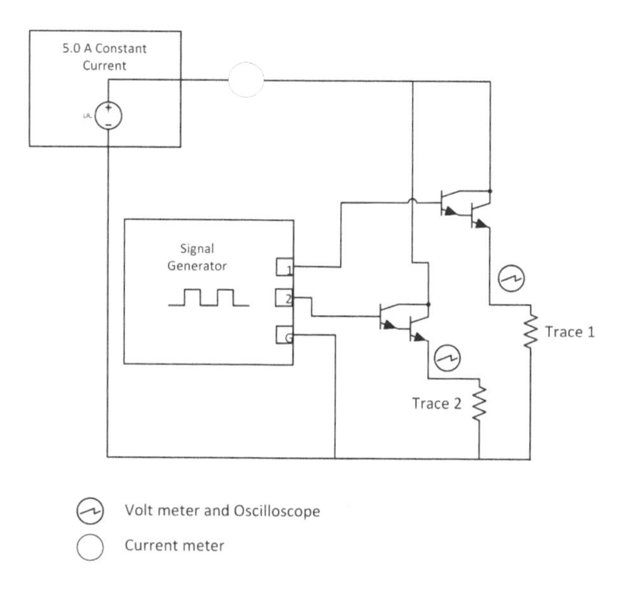

Figure 13.7
Diagram of experimental test setup.

For the first test, the duty cycle was set to 50% and the frequency varied from 0.05 to 10,000 Hz. The results are shown in Figure 13.8. The temperature is constant at 45 degrees C within the resolution of the thermocouple. This compares to the 45.4 degree C temperature recorded in Table 13.1.

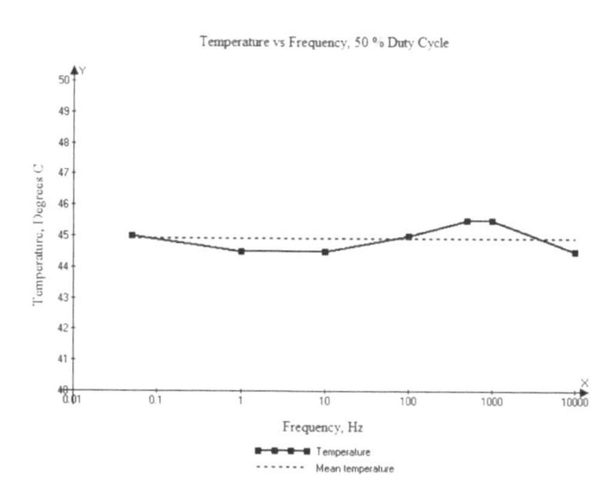

Figure 13.8
Temperature at 50% duty cycle as a function of frequency.

Next, we looked at the temperature as a function of duty cycle at two different frequencies, 100 Hz, and 10 KHz. Those results are plotted in Figure 13.9. They reflect a linear relationship, within the probable experimental error. These results are consistent with the conclusions reached above with the TRM model.

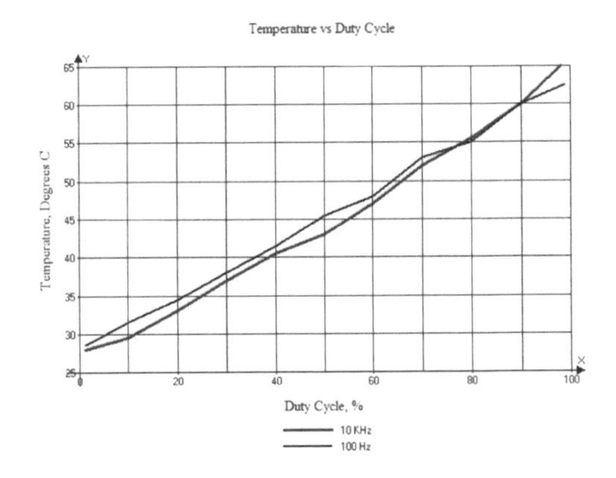

Figure 13.9
Temperature as a function of duty cycle.

Finally, we recorded the actual temperature for a 0.04 Hz frequency at an 80% duty cycle (after the temperature cycles had had time to stabilize). The results are shown in Figure 13.10. We have also included some data from the TRM PWM analysis (Figure 13.5) plotted on the same axes for the same conditions for comparison. Recall that the thermocouple has a 0.5 degree resolution and a 1 second sampling rate. The curve also reflects some "wiggles" caused by the difficulty in holding the thermocouple steady for several seconds time! Also recall that the PWM analysis slightly understates actual because it does not take the thermal coefficient of resistivity into consideration.

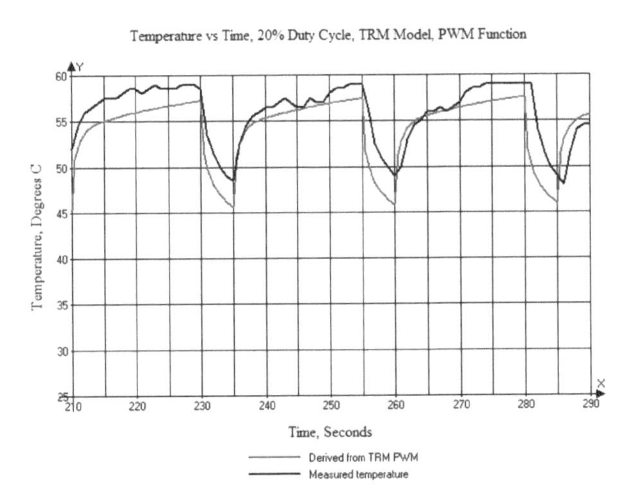

Figure 13.10
Temperature vs time for a 0.04 Hz, 80% duty cycle waveform.

13.2.1 Conclusions: It would appear that we can confirm the preliminary conclusions reached above:

1. The temperature of a trace is determined by the equivalent RMS value of the current driving it.
2. The temperature of a trace at any frequency and duty cycle seems to be independent of frequency (at least at frequencies below where the skin effect comes into play.)

13.3 Analog AC Currents

But a question still remains: What about analog ac currents? What about sine or triangular waveforms? The next section looks at the slightly more complicated topic of analog waveforms to see if the same conclusions apply.

13.3.1 Test Circuit: The first problem is how to apply an analog signal of known, repeatable current level, through a trace at a significant enough current to heat the trace? To accomplish this, the circuit shown in Figure 13.11 was created. A waveform generator applied one of three signal types to the base of the Darlington transistor. That resulted in a current through the trace controlled by the current through the collector. That current was supplied by a 10.0 volt constant voltage generator through a 1.05 Ohm resistor (with a very low temperature coefficient.) The current through the trace is derived as:

Current = (10.0 - voltage at collector)/1.05

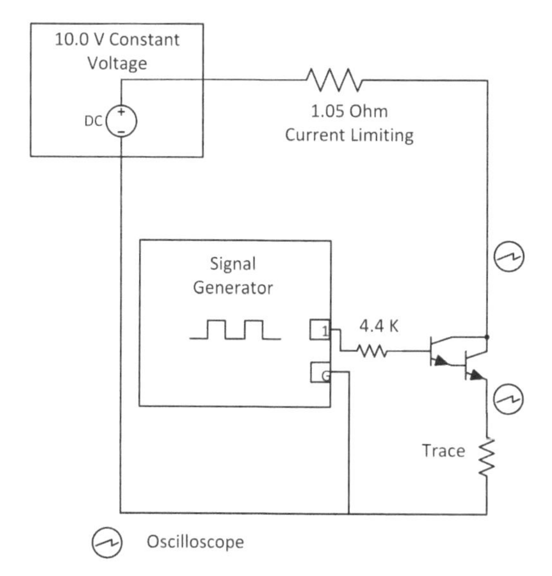

Figure 13.11
Test schematic

The waveform generator settings were extremely flexible. The output voltage could be set from 0.0 to 10.0 volts. The frequency could be set from 0.01 Hz to over 20 MHz. And the signal DC offset could be set from -10.0 to +10.0 volts DC. Three waveforms were employed during this analysis; sine, triangle (actually sawtooth) and square. The duty cycle for the square and triangular waveforms could be set from 0.0 to 99.9 percent. The duty cycle for the triangular waveform was set at 99.9 percent (sawtooth), and the duty cycles for the square waves were set at 50%, 75% and 99%. Output levels were set to result in large output currents, but not large enough to switch the transistor off nor to bring it into saturation. Samples of the waveform generator output waveforms are shown in Figure 13.12. The generator waveforms exhibit no measurable distortion.

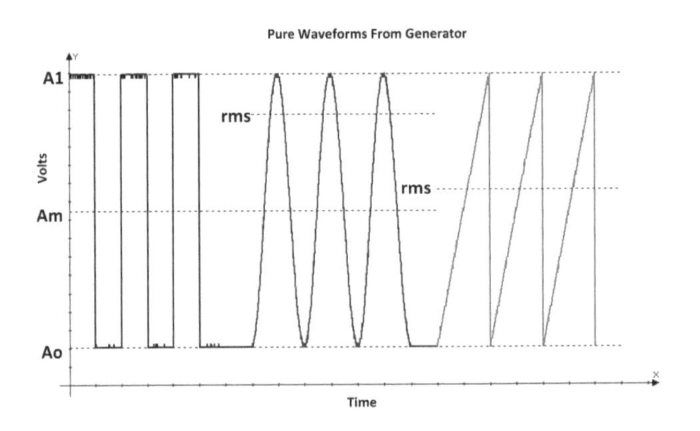

Figure 13.12
Waveform generator outputs.

13.3.2 RMS Signal Levels: Next, we need to consider how to calculate the rms values of the signals [6]. There are some reference lines drawn on Figure 13.12:

A1	Peak voltage
A0	Minimum voltage
Am	Average voltage; i.e. Am = (A1-A0)/2)

If we ignore any dc offset, the rms value of a simple square wave is simply its peak voltage, shown as (A1-Am) in Figure 13.12. But if the square wave has a duty cycle other than 50%, it looks more like the pulse train in the previous sections. The rms value of the pulse train is its peak value multiplied by the square root of the duty cycle, or (A1-Ao)*SQRT(duty cycle). Similarly, if we ignore any dc offset, the rms voltage of a simple sine wave is .707*peak, or .707*(A1-Am) in Figure 13.12. And, ignoring any dc offset, the rms value of a simple triangular (or sawtooth) waveform is 0.577*peak, or .577*(A1-A0) in Figure 13.12.

If there is a dc offset, the rms values are calculated as follows:

Let rms = the rms value of the simple waveform.

Let dc = the dc offset (Am for the simple square wave and the sine wave, Ao for the sawtooth wave and the pulse train.)

Then the rms value of the offset waveform is: $rms_{offset} = \sqrt{rms^2 + dc^2}$

13.3.3 Non-Linearities: But there can be a problem in a practical circuit. The circuit (transistor) is non-linear over the range of current applied here. For example, Figure 13.13 shows the resulting sawtooth waveform. The degree of non-linearity is obvious.

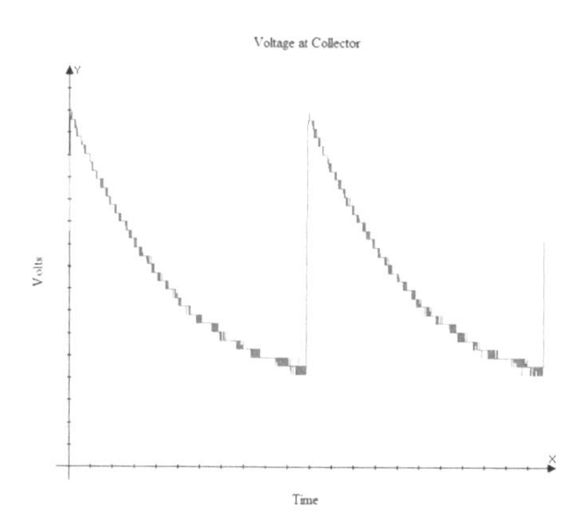

Figure 13.13
Sawtooth voltage waveform at the collector vs time

The oscilloscope used in this analysis [7] is a digital scope whose output can be saved to a computer file. The scope also has the ability to calculate the "True RMS" voltage (as well as the dc average, and many other measures) of the waveform under evaluation. Therefore, we are freed from having to manually calculate the rms value.

But then, there is another problem. In our circuit, high voltages at the collector correspond to low currents through the trace, and low voltages at the collector correspond with high currents through the trace. The root-mean-square calculation tends to bias the results to the higher values, in this case the opposite of what we want. Fortunately, the scope can output a data file which can be imported into an Excel spreadsheet. The data can be corrected and the rms value calculated by the spreadsheet [8].

13.3.4 Results: The trace used in this analysis was one that was on the same board as one that used in our fusing investigation (Section 11.3.) It was a 100 mil wide, 0.5 Oz, 6" long trace. The resulting summary data are shown in Table 13.3 and graphed in Figure 13.14. These results were obtained at a frequency of 100 Hz [9]. The black line in Figure 13.14 is the current/temperature relationship for the trace with constant DC current applied. These results are consistent with the prior results, i.e. the effective current for current/temperature analyses of ac waveforms is the rms value of the current.

	Peak V	Trough V	RMS current (A)	Trace Temp °C
Sine	7.5	2.1	5.39	65.5
Sawtooth	7.15	2.0	5.68	68.5
Sq, 50%	7.0	2.25	5.11	62.0
Sq, 75%	7.68	2.25	6.12	75.0
Sq. 99%	7.9	2.25	7.27	97.0

Table 13.3
Summary results of testing.

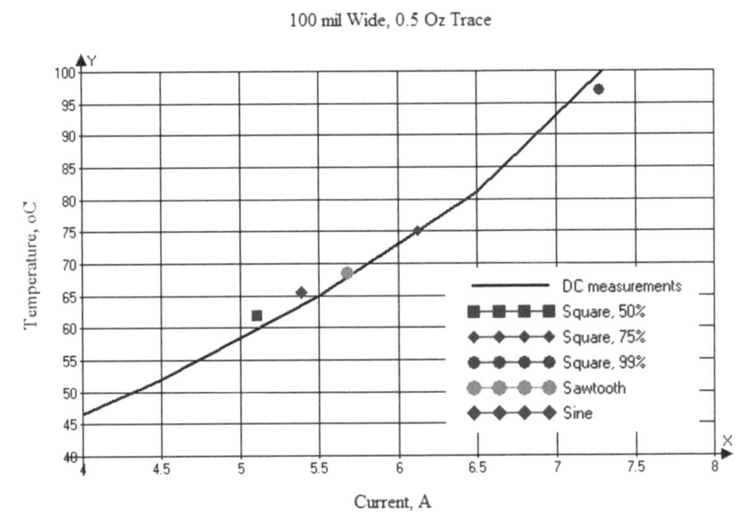

Figure 13.14
Graph of summary results.

When applying the signals to the trace, we also varied the frequency from 0.5 Hz to 1,000 Hz. The temperatures remained constant with frequency within the measurement accuracy of the study (as expected). (At frequencies significantly above 5,000 Hz our test setup became unstable.)

13.3.5 Conclusion: Again, we conclude that the results of these analyses are consistent with the prior results:

1. The effective current of an analog ac waveform is the RMS (root mean square) current, and
2. The temperature of an analog ac waveform is independent of frequency (at least over the tested range.)

Notes:

1. See Chapter 6.2.
2. As the trace temperature increases, the resistance of the trace increases (because of the thermal coefficient of resistivity.) Thus the trace temperature increases even higher. TRM can take this into account in its calculations, as shown in previous sections. However, this capability has not yet been programmed into the PWM module (discussed a little later in this section). Therefore, the PWM module tends to understate the temperature slightly. That effect is noticeable, but not large, in the temperature ranges investigated in this chapter.
3. The slight "glitches" in the curve are caused by adjustments to the HTC value in the model, see Section 6.1. The curve should be perfectly linear.
4. Applying the RMS value for a 50% duty cycle for the first 60 seconds gets the curve "up to temperature" a little more rapidly.
5. The trace is a 27 mil wide, 2.9 mil thick, and 5.9 inch long. It will have a resistance at 60 degrees of approximately .063 Ohms if it is uniformly heated. I^2R for 5 Amps would therefore be approximately 1.58 W. But the trace is not uniformly heated, it cools towards the pads at each end. The model calculates that the total power dissipated in the trace is 1.464 Watts.
6. See https://en.wikipedia.org/wiki/Root_mean_square
7. The oscilloscope used is a PicoScope model 2204A.
8. There are approximately 900 data samples per cycle in our waveforms. The Excel formula for calculating rms is =SQRT(SUMSQ(A1:A10)/COUNTA (A1:A10)) where A1:A10 is the data range of interest.
9. At frequencies above about 10 Hz the thermal inertia is such that the temperature does not change much within the cycle.

14 Right Angle Corners

14.1 Background:

Through the years the impact of right angle corners on printed circuit boards (PCB) traces has been a hot topic. As early as the 1990s some people were arguing strongly against the use of 90° corners, using mitered corners instead. They based this recommendation, typically, on one of two reasons:

A. The width of a typical PCB trace increases as the trace turns a corner. This increases the local capacitance at that point, lowering the trace impedance. This impedance discontinuity can therefore set up a reflection that can lead to signal integrity issues.

B. The current density around a trace corner shifts so that it is strongest at the inside edge of the corner where the path length is the shortest. This increased current density can lead to an increase of the local electromagnetic field, which can subsequently lead to increased EMI from the trace.

In a landmark paper published in 1998 [1], Brooks tried to put these arguments to rest. In that paper he showed:

A. Although it is true the width increases, it does not increase very much. Furthermore, the maximum width occurs *at a point*, not over a range. And finally the entire signal path around the corner is *much* shorter than the "critical length [2]." Therefore the impedance discontinuity is insignificant.

B. An experimental investigation of actual EMI measured from traces with various corner designs, including a diabolical 135 degree corner with sharp "points' (which would never be found on a PCB!) showed no significant radiation greater than a control "straight" trace [3].

A more recent paper [4] shows an image (reproduced as Figure 14.1) of a 10 mm (400 mil) wide, 35 µm (1.0 Oz.) thick trace, carrying 18 Amp, and heated to approximately 80° C. The image suggests that there is about a three to four degree C difference between the outer edge of the trace and the inside corner. The authors conclude:

This corner effect decreases the CCC (current carrying capacity) of bended traces compared to straight trace with the same width….Value will be added if CCC design chart for bended traces can be a part of future PCB trace thermal design guideline.

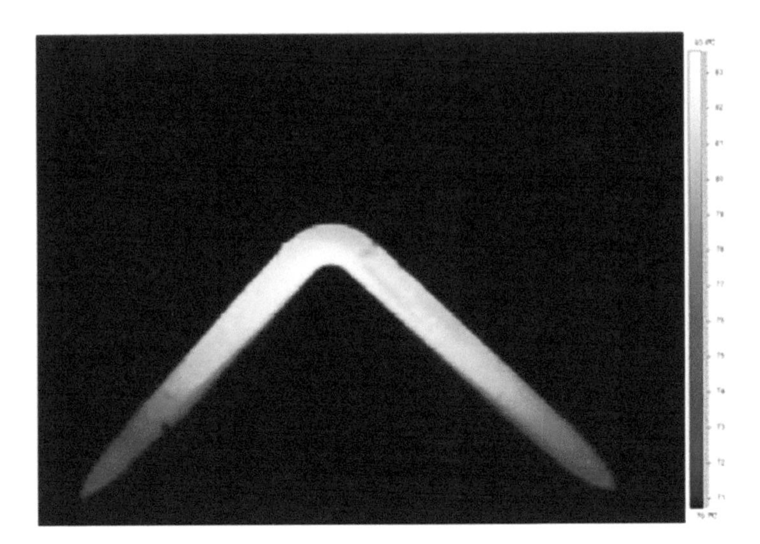

Figure 14.1
Thermal effect around right angle corner, Source: Wang, et al [4].

This chapter looks at the thermal gradients around the corners on a PCB.

14.2 Trace Heating and Cooling Dynamics:

PCB traces heat by I^2R power dissipation in the trace. Traces cool by conduction through the board material, convection through the air, and by radiation. Temperature stability occurs when the rate of heating equals the rate of cooling.

As shown in previous chapters, conduction through the board material is the predominant cooling mechanism on a typical PCB. Recall that conductive cooling occurs in 2 directions: "in-plane" (parallel to the traces) and "through-plane" (through the board perpendicular to the traces.) "In-plane" conduction is the predominant mechanism of the two.

We have shown in Chapter 12 that there is a thermal gradient across the trace, hottest along the centerline and cooling out towards the edge of the trace. This is because of "in-plane" conduction and the fact that the cooling path is shorter for heating at the edge of the trace than it is at the centerline of the trace. This fact will be important in the discussion below.

14.2.1 Software Simulation: We used TRM to look at traces with right angle corners. Four corner configurations were looked at, illustrated in Figure 14.2.

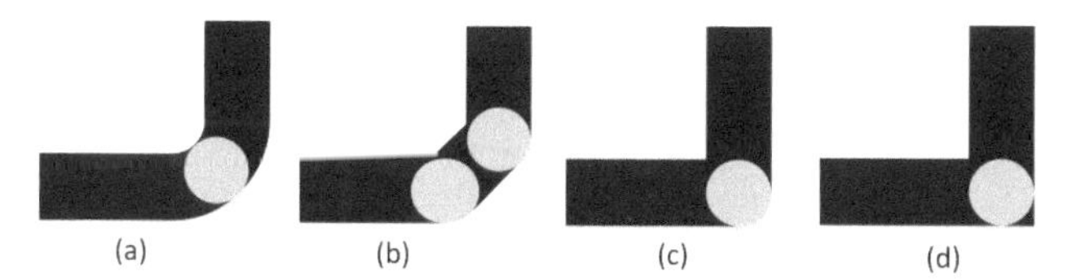

Figure 14.2
Four different corner configuration

Configuration (a) is a constant-width corner, i.e. the trace is the same width everywhere. It has a rounded inner corner. Configuration (b) is typical of a trace found on a PCB. It has mitered 45 degree corners whose outer diameter is the same as the trace width. Configuration (c) is also typical of what you might see on a PCB. It is typically created by "drawing" a trace with a circle (whose diameter is the same as the trace width). Configuration (d) is a "square" corner, rarely seen on a PCB, and actually somewhat difficult to create on a PCB (it can be drawn with a square aperture on some plotters). It should be noted that the width across the corner, and therefore the trace cross-sectional area at the corner, is the same as the body of the trace for configuration (a), but increases as we move to configuration (b) and then on to configuration (d).

The basic model we used was a 5 mm (200 mil) wide, 35 µm (1.0 Oz.) trace on a standard 1.5 mm thick FR4 substrate. All four trace configurations were modeled with a current of 8.0 A, resulting in a temperature of approximate 44° C (about a 24° C increase.) The TRM model allows us to investigate (among other things) both the current density and the thermal characteristics around the corner.

Figure 14.3 shows the current density (A/mm^2) around the corners of the four configurations. There is a fairly dramatic shift in current from the outside corner to the inside corner, especially for configurations (c) and (d), where the inside corners are sharp [5]. This is directly related to the path length of the current around the corner, that path being significantly shorter along the inside edge.

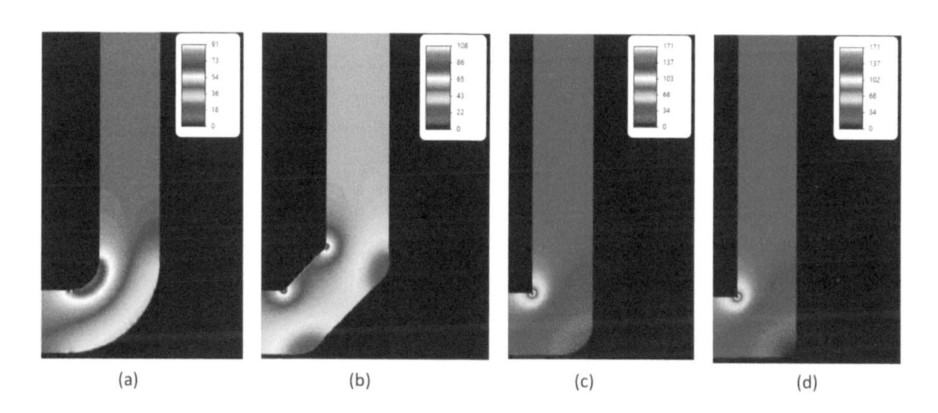

Figure 14.3
Current density around the corners of the four trace configurations shown in Figure 14.2, each carrying 8.0 A.

One might speculate that the increased current density at the inside edge of the corner would lead to a temperature increase along the trace at that spot. Another argument, however, might be that the thermal conductivity of the copper is so high that no temperature gradients could develop. Figure 14.4 shows the thermal profiles for the four configurations modeled. The small boxes record the temperature at the upper left corner of the box.

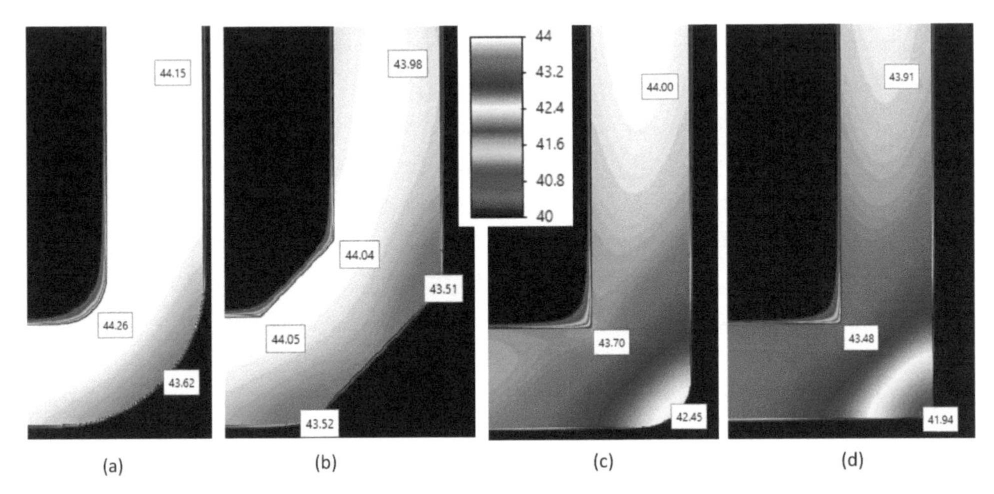

Figure 14.4
Thermal profile around the corners of the four trace configurations.
The temperature range shown is between about 40° C to 44° C

The models suggests that the trace temperature is slightly warmer at the inside edge of the configurations than at the outside edge. The temperature differences range from about 0.5 degrees C to about 1.5 degrees C, and are larger the greater the width at the corner exceeds the width of the trace. Interestingly, however, the inside corner is either about the same temperature as the basic trace or is cooler than the basic trace. This is because the increased width at the corner offers a lower resistance to the current through the corner and therefore a lower I^2R heating around the corner. [See note [6].)

But it turns out the heating dynamics around the corners are much more complex than this. Figure 14.5 illustrates the thermal profiles of trace configurations (a) and (c) (the constant width corner and the "drawn" corner) and of the board around the traces. Note that the board around the inside of the corners has a wider thermal profile than around the outside of the corner. This indicates the outside of the corner is cooling more efficiently than is the inside of the corner. There are at least four factors that contribute to the thermal profile of the trace and the surrounding area:

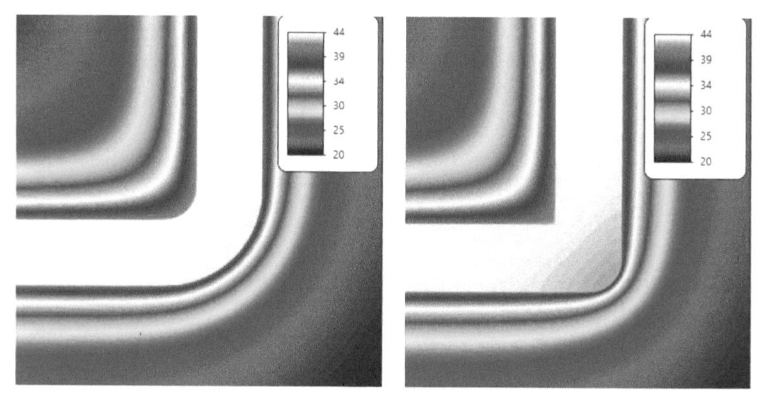

Figure 14.5
Thermal profile of the modeled traces and the board area around the traces.

A. If the cross-sectional area of the corner is larger than the cross-sectional area of the rest of the trace, the "point" resistance around the corner will be lower than the "point" resistance along the rest of the trace. Therefore, the temperatures around the corner will tend to be smaller than they are along the rest of the trace.

B. The current density is significantly higher toward the inside of the corner than it is toward the outside of the corner. Therefore, the "point" I^2R power dissipation will be higher at the inside of the corner than it is at the outside of the corner.

C. The thermal conductivity of the copper trace is so high that some of the heat generated at the inside corner will conduct away through the copper to the outside corner.

D. Notice how the *board* area at the inside of the corner tends to be much warmer than the *board* area at the outside of the corner. This is because the heat from both legs of the trace conducts into the same 90° arc of the board on the inside of the corner, but the heat tends to fan out (in a 270° arc) away from the outside of the corner. Thus, the outside of the corner cools more effectively than does the inside of the corner. It is likely that this is the most important determinant of the of the trace temperature around the corner.

Now, looking back at Figure 14.4, Configuration (c) (Drawn Corner), it is true that the inside of the corner is warmer than the outside of the corner, it is *also* true that the inside edge is *cooler* than the midpoint of the straight section of the trace on either end of the corner. This illustrates the impact of the complex relationships outlined in A through D above.

Additional modeling, not reported here, results in the following observations:

1. The greater the distance from the inside corner to the outside edge of the curve, the more pronounced is the difference in temperature.

2. The sharper the inside corner, the more pronounced is the difference in temperature. (That is, smooth inside curves result in lower temperature differences around the corner.)

3. Narrower traces and/or lower currents result in smaller temperature differences around the corner.

14.3 Experimental Verification [7]

The experimental investigation was performed by Norocel Codreanu with a FLIR SC640 infrared (IR) camera. The electromagnetic radiation in the 7.5 - 13 µm range hits the active layer and changes its electrical resistance. This resistance change is measured, processed and transformed by the IR camera into a temperature array which can be represented graphically.

The test board (Figure 14.6) was designed using Cadence/OrCAD 16.6-2015 Electronic Design Automation (EDA) software system and produced in the UPB-CETTI labs using PnP prototyping technology.

Figure 14.6
Test board.

The PCB laminate of the test board has a standard FR-4 substrate, 1.5 mm (60 mil) thick with a copper thickness of 35µm (1.0 Oz.). The layout consists in two areas of PCB structures, placed in the upper and in the lower areas of the board, having in the middle the horizontal symmetry line. The lower area is identical, but mirrored vertically, with the upper area. Each area is composed of:

(a) On the left side: one PCB trace of 5 mm in width (drawn-trace) with longer miter, one PCB trace of 5 mm in width (drawn-trace), with shorter miter, and one PCB trace of 5 mm in width (constant-width), with fillet/round corner;

(b) On the right side: one PCB trace of 0.5 mm in width (drawn-trace) with right angle corner (90°), one PCB trace of 2.5 mm in width (drawn-trace) with right angle corner, one PCB trace of 5 mm in width (drawn-trace) with right angle corner, and one PCB trace of 5 mm in width (square-corner, see figure 14.2, configuration (d)) with right angle corner.

A current of 8 Amps was applied to the test traces and then photographed with the thermal imager. Figure 14.7 shows the thermal pattern for the drawn corner (left) and the square corner (right). Compare these results with those shown in Figure 14.4 configurations (c) and (d). The thermal images compare favorably with the modeled results.

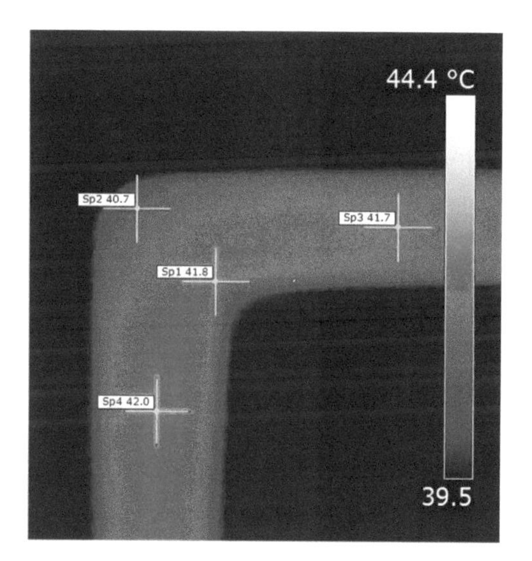

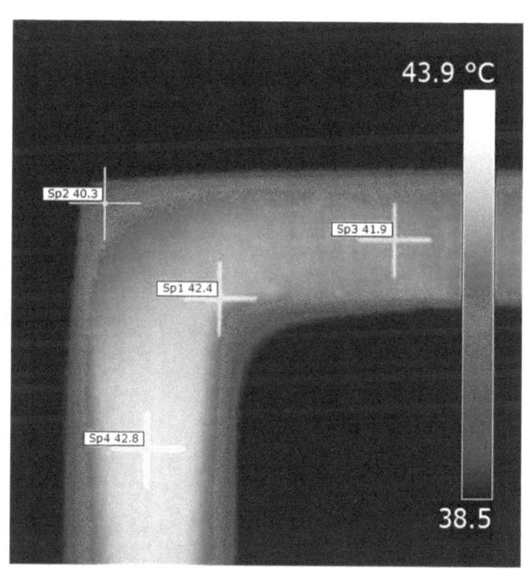

Figure 14.7
Thermal profiles of the test board: drawn corner (left) and square corner (right)

Figure 14.8 illustrates the thermal profile of the same traces as shown in Figure 14.7 along with the surrounding board area. Compare these images with those shown in Figure 14.5. Again the two results compare favorably. This illustrates how the outside of the traces cool more effectively than do the inside of the traces.

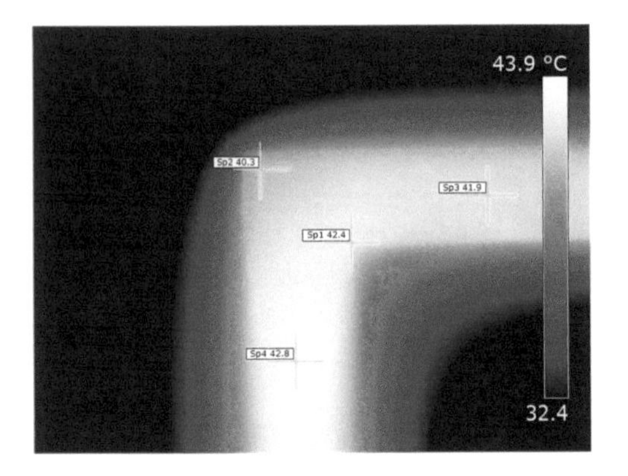

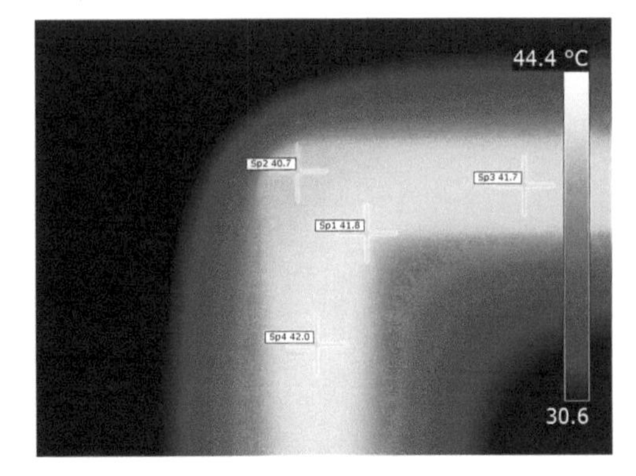

Figure 14.8
Thermal profiles of the test board and surrounding board area:
drawn corner (left) and square corner (right)

14.4 Conclusions:

Current densities around corners are higher at the inside of the corner compared to the outside of the corner. This is consistent with previous studies.

Temperatures are higher at the inside of a corner than they are at the outside of a corner. And there is a close correlation between the modeled data and the experimental data. But these temperatures are not necessarily higher than they are along the main body of the trace. This is because the traces are usually wider at the corner than they are along the main body, and the electrical resistance is therefore lower. Consequently the I^2R heating is lower at that point.

The temperature imbalances around a corner are not likely the result of the difference in current densities. More likely they are the result of the significantly more efficient cooling mechanism around the outside of the corners where the open board area is much wider.

Notes:

1. Douglas G. Brooks, "90 Degree Corners: The Final Turn," Printed Circuit Design Magazine, January 1998, available at http://www.ultracad.com.

2. For a discussion of critical length, see Douglas Brooks, PCB Currents; How They Flow, How They React, Prentice Hall, 2013, p. 216

3. While it is true that there are no problems with using right angle corners on traces, a great many designers still prefer mitered corners for aesthetic reasons. Some fabricators prefer mitered corners to right angle corners for fabrication reasons.

4. Yi Wang, S. W. H. DeHaan, J. A. Ferreira, "Thermal Design Guideline of PCB traces under DC and AC Current," Energy Conversion Congress and Exposition, 2009. ECCE 2009. IEEE.

5. If a model has a sharp, 90 degree corner, the mathematics may result in an infinite current density at that specific (infinitely small) point because of a singularity (think divide by zero). This does not have any effect on the rest of the model.

6. We also modeled Wang's et all 10 mm trace, heated to near 80 degrees C. The model does confirm Wang's et al three to four degree difference between the outside and inside corners for that trace size heated to that temperature.

7. I am indebted to Norocel Codreanu, Ph.D., who conducted the experiments and allowed us to use these images. Dr. Codreanu is full professor at "Politehnica" University of Bucharest (UPB), Romania, Faculty of Electronics, Telecommunications and Information Technology, and currently the executive manager of the UPB university research center "Center for Technological Electronics and Interconnection Techniques" (UPB-CETTI).

15 INDUSTRIAL (CT) SCANNING

This chapter would not have been possible without the generous support of Fabio Visentin and Alejandro Golob of Jesse Garant Metrology Center, Windsor, Ontario. I sent them a portion of a test board and they provided some scanning data of that board to review.

15.1 Scanning Process

Industrial CT Scanning (also sometimes called Computed Tomography) [1] is somewhat similar to a medical CAT scan. Referring to Figure 15.1, there is an x-ray source that sends out a beam in a roughly conical shape. The target (think PCB) is placed in the path of the beam. A sensor collects the beam and generates a digital representation of the target. The closer the target is to the source, the larger will be the magnification. But, the closer the target is to the source, the smaller is the area of the target that can be analyzed.

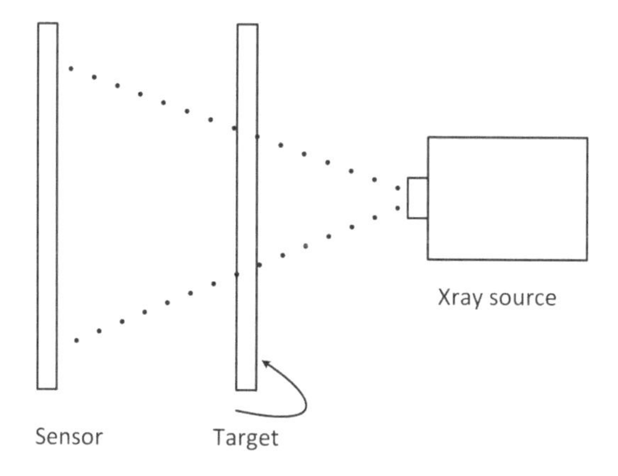

Figure 15.1
X-ray process.

The target is rotated 360 degrees. Individual x-rays are taken as the target rotates. Roughly 2,200 x-rays are taken during a 360 degree rotation! The magic happens when the software collects all those digital images and combines them into a data file that can be viewed as a 3-dimensional image (with a special software viewer). Figure 15.2 illustrates the actual equipment used for generating the x-rays for this project.

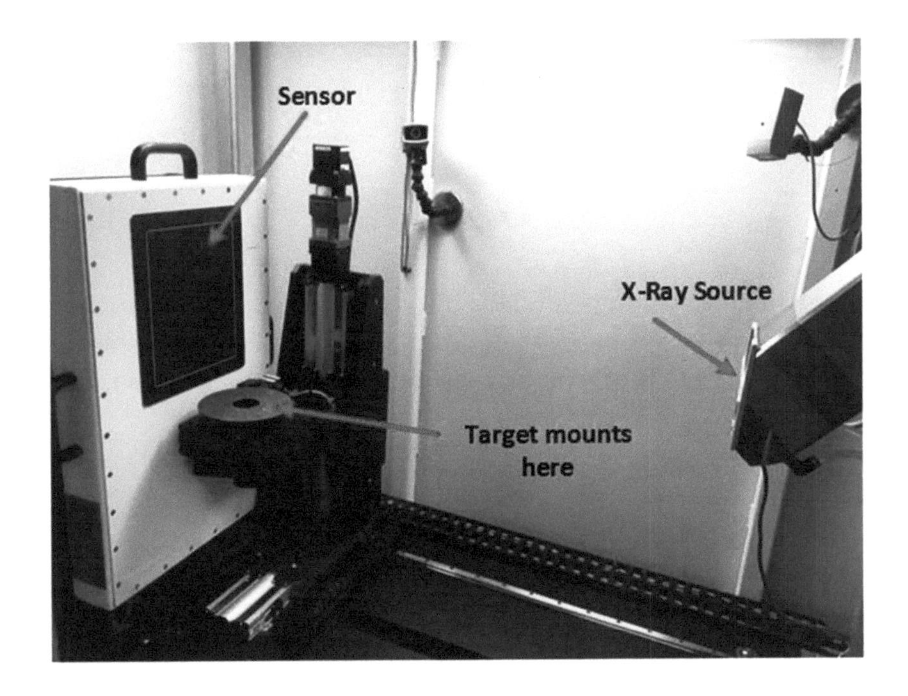

Figure 15.2
X-ray equipment used for this project
(Courtesy Jesse Garant Metrology Center.)

The view that got me excited about computed tomography is shown in Figure 15.3. This is from one of the libraries at Jesse Garant Metrology Center. What is readily apparent is that a trace is missing at one of the pins on one of the inner layers. (The desire to locate the exact point of this failure was one of the reasons this particular board was x-rayed in the first place.) It is possible, by adjusting the depth of this image, to actually look inside the IC packages mounted on top the board.

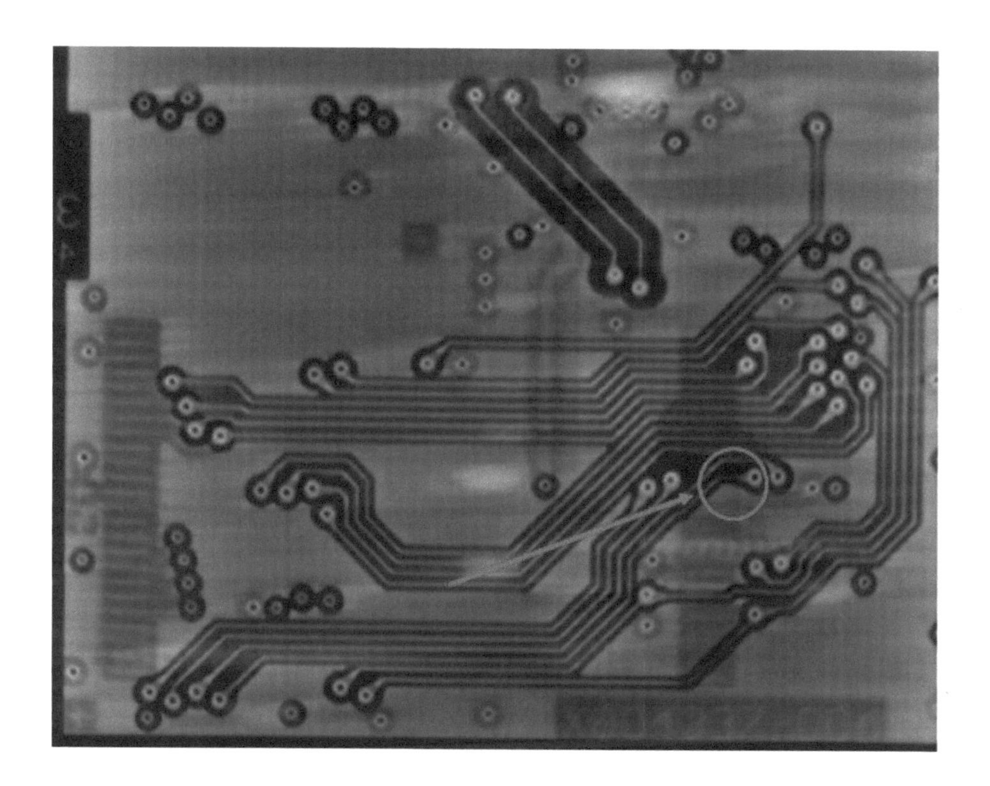

Figure 15.3
View of an inner layer of a board. Note the failure at the point indicated by the arrow.

15.2 Results

Figure 15.4 shows my particular test board. The viewer is positioned at a point right at a via. The location is shown by the crossed lines and red plane indication in the view at the bottom right. The bottom right image is that of the board (it was cut down to fit on the target stand. More about that later.) The other three quadrants provide views in the three primary axes. The upper right quadrant is viewing down onto an x-y plane at a particular point along the z-axis (y is up; x is to the right; z is into the page). In this view, the plane is the middle of the trace slicing through the via. The line points to the via. The upper left quadrant is the view looking back in the y-direction at a point along a plane in the x-axis (y is up; z is to the right; x is into the page). The line points to the same spot on the via. The lower left quadrant is looking from left to right on the x-axis along plane in the y-axis (x is up, z is to the right). Again, the line is pointing to the same spot on the via.

The image point can be moved to any arbitrary point in 3-dimensional space and can be zoomed in or out to reveal any desired area. One of the many tools available in the viewer is the precise measurement between any two points in the board. Unfortunately, the resolution in this image is not quite sharp enough to measure the via width or the thickness of the copper plating.

Two aspects of this ability are particularly intriguing:

1. I can look at any point inside any and all vias,
2. I can examine the entire length of any trace in all three dimensions.

But the problem? The resolution is not quite good enough to take measurements with the required accuracy. The process is excellent for qualitatively analyzing the makeup of the board (including the interior of any components that might be mounted on the board), but not quite suitable for quantitatively measuring the trace parameters. So we can tell if a trace is missing, or if there is a marked difference in the plating thickness on opposite walls of a via. Within reason, we can locate opens and shorts (except possibly hairline shorts). We can evaluate if there are significant non-uniformities in the plating along a trace. We can even evaluate some solder problems. But we can't precisely measure any of those anomalies.

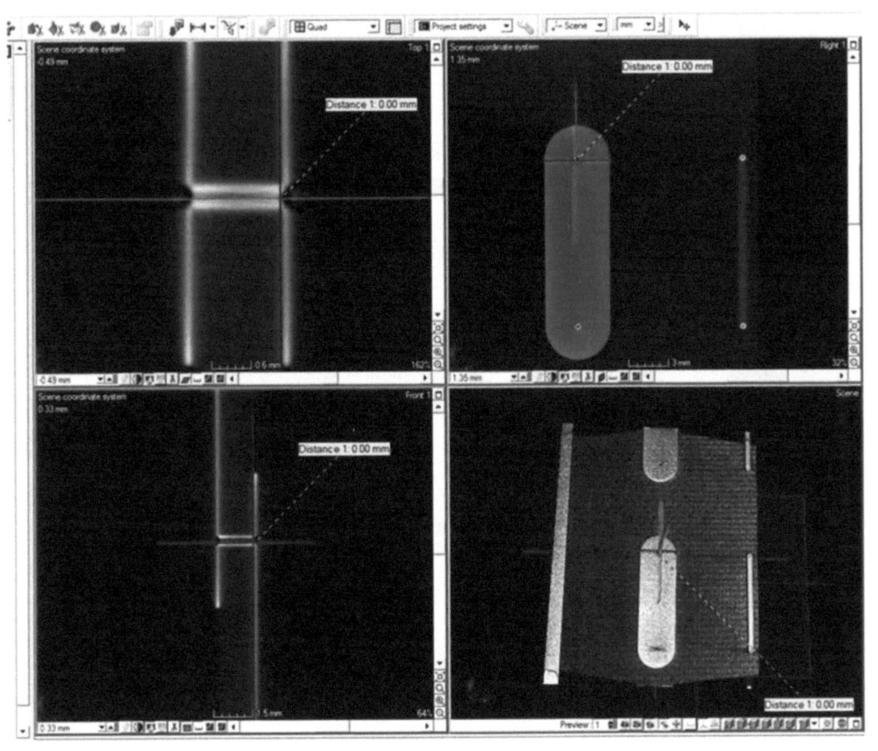

Figure 15.4
x-ray images of my board

Can the resolution be improved? Perhaps. If the target is moved closer to the x-ray source, then the image is larger and more precise measurements are possible. But a smaller area of the total board is then visible. Here are some possible advances:

1. Increase the power at the source (along with some attendant power management and cost issues.)
2. Shorten the wavelength of thc source.
3. Increase the sensitivity of the sensor.

All of these are potential solutions, but they are not practical today. We can anticipate that one or more of them might become practical at some point in the future.

Notes:

1. There are some significant differences between Industrial CT Scanning and a CAT scan. The output of Industrial CT Scanning is a data file that allows the user to look anywhere in 3D space. The output of a CAT scan is typically one or more 2D views of the area of interest in the patient. A simple medical x-ray of a patient produces a single picture. Single x-rays of an industrial product is referred to as Industrial Radiography. Industrial Radiography and Industrial CT Scanning can be done at much higher power levels than can be done in medical, which means they can scan denser materials with higher resolution than is available in medical imaging.

Part 4

APPENDICES

A1 MEASURING THERMAL CONDUCTIVITY

In order to run accurate simulations, as described in Chapter 6, one of the necessary parameters is the thermal conductivity coefficient of the PCB dielectric material. Recall from Section 4.2 that PCB materials are anisotropic. That is, they cool better in the in-plane direction than they do in the through-plane direction. We need to know the thermal conductivity coefficients in both directions for reliable modeling.

Data sheets frequently ignore this parameter. If they do include a thermal conductivity coefficient, they frequently only include a single value, and they frequently fail to state which direction the coefficient applies to.

A search on the Web provides anecdotal estimates of PCB material thermal conductivity ranging from 0.3 to 0.8 W/mK (Watts per meter degree Kelvin). Wikipedia references at least six different ways to measure thermal conductivity [1]. The people at C-Therm Technologies [2] in New Brunswick, Canada, were nice enough to support this research effort by measuring the thermal conductivity coefficient of the board material used in Chapter 6.

C-Therm uses a technology known as the Modified Transient Plane Source (MTPS) technique. The process involves a transducer (see Figure A1.1) that provides a constant current heat source for the sample. The sample (approximately 1.5" square) is placed on top of the sensor. A short burst of heat (for perhaps a second or two) starts to heat the sample.

If the sample is very thermally conductive, the heat will drain away from the sensor through the sample, lowering the temperature at the sensor. If the sample does not conduct heat well (i.e. has a low thermal conductivity), then the sensor will be hotter. Thus the temperature at the sensor is inversely proportional to the thermal conductivity of the sample.

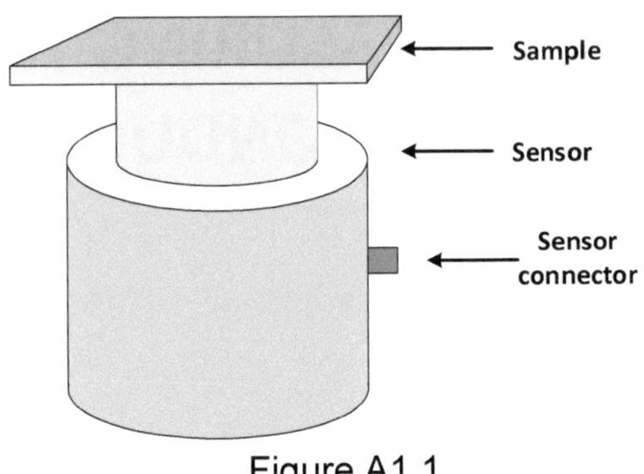

Figure A1.1
MTPS process.

If the sample is placed on the sensor as shown in Figure A1.1, the process measures through plane (z-axis) thermal conductivity. To measure in-plane thermal conductivity, multiple samples are grouped together (and compressed for good thermal conductivity) edge-wise on the sensor. Figure A1.2 shows the actual through-plane and in-plane measurements of the samples from the materials used in the simulations described in Chapter 6. In Figure A1.2 the actual sensor is placed on top of the sample, weighting it down (and thereby making good thermal contact.) The base provides a good, controlled heat sink for the measurement process.

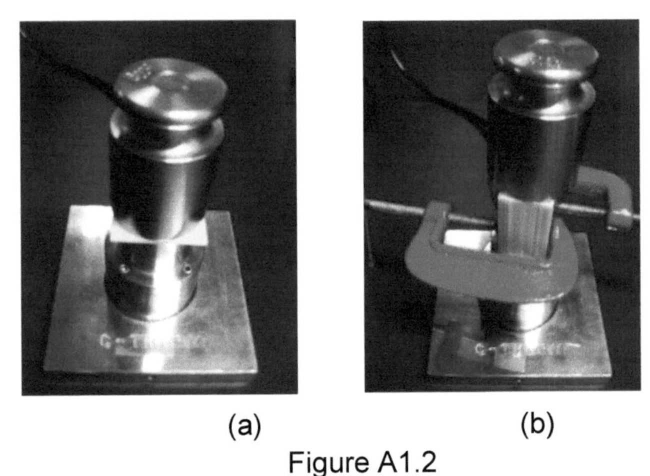

(a) (b)

Figure A1.2
Through-plane (a) and in-plane (b) measurements of thermal conductivity for the material employed in the simulations described in Chapter 6.

Notes:

1. https://en.wikipedia.org/wiki/Thermal_conductivity_measurement
2. http://www.ctherm.com/

A2 MEASURING RESISTIVITY

There are numerous anecdotal references to measurements of trace resistivity on the Web with results ranging from about 1.7 uOhm-cm to 2.4 uOhm-cm. In my opinion most of them suffer from various forms of measurement error. That is why I prepared this short reference on how to measure resistivity.

A2.1: Resistance vs. Resistivity

First, *resistivity* is a parameter associated with an element (such as copper.) *Resistance* is a parameter associated with (in our context) a trace. These are different things. The relationship between resistivity and resistance involves the cross-sectional area of the trace and its length and is given by Equation 3.1 (repeated here as Equation A2.1.

[Eq. A2.1] $R = (\rho/A)*L$

Where R = resistance in Ohms
 ρ = resistivity in Ohms-length
 A = cross-sectional area in square units
 L = length in units

So, for example, if we change the trace length or cross-sectional area, the resistance will change, but the resistivity of the copper will not.

The resistivity of pure copper is usually given as 1.676 uOhm-cm at 20° C. The resistivity of copper alloys can vary, but it is usually given as 1.72 or so. By comparison, silver has the lowest resistivity of any element at 1.6 uOhm-cm. So any reported measurement of copper resistivity less than 1.6 uOhm-cm (at 20° C) is impossible! There is no element or combination of elements that has this low a resistivity (unless you lower the temperature.)

There are anecdotal reports of resistivity measurements of plated copper well above 1.7 uOhm-cm (at 20° C) but these results should be suspect because plated copper is *pure* copper, almost by definition. The plating process electro-deposits copper ions (and no other ions) on the cathode. Therefore, plated copper should result in a measured resistivity close to that of pure copper.

A2.2: How to Measure PCB Trace Resistivity

The practical way to measure trace resistivity is to start by measuring trace resistance. Then use Equation A2.1 to calculate the resistivity. There are two practical ways to measure trace resistance. One is to apply a known current to the trace and measure the voltage across the trace (or vice versa). The other is to use a very sensitive Ohmmeter. I will show you that these are actually the same thing!

If you want to measure the resistivity (in fact if you want to measure anything), it is usually beneficial to use the largest parameters practical. By that I mean you want the largest voltage and current practical. You want the largest cross-sectional area practical. And you want the longest length practical. The reason is that it is generally true that larger parameters can be measured more precisely than smaller parameters can. You can usually measure inches to a finer number of decimal places than you can measure mils. You can usually measure hundreds of mv more precisely than you can measure tens of mv.

And then, where and how you place your probes is very important in measuring resistance. Figures A2.1 and A2.2 show two different probe attachment methods for applying a current to a trace and measuring the voltage across the trace.

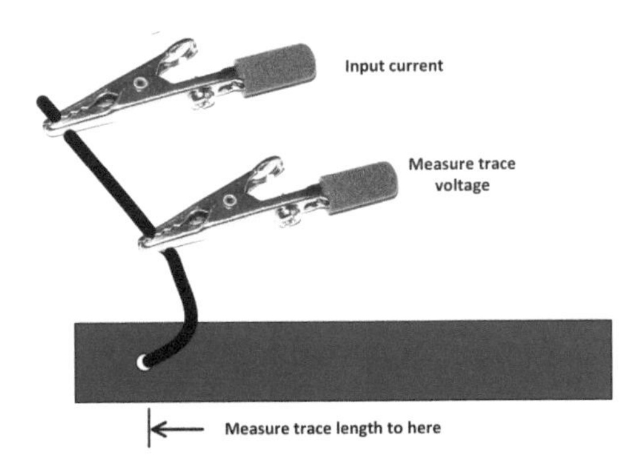

Figure A2.1
Incorrect method for measuring trace resistance.

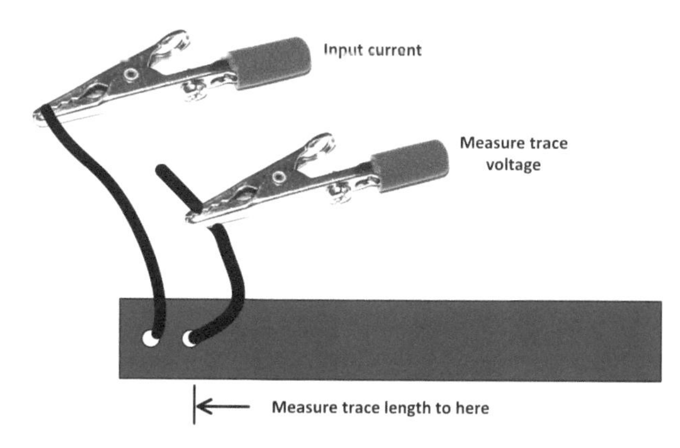

Figure A2.2
Correct method for measuring trace resistance.

The reason Figure A2.1 is wrong is because the measurement of the voltage across the trace includes the section of the wire lead carrying the current between the lower connection point and the trace. In effect, you are measuring the voltage across the trace, part of the wiring, and at the contact resistance between the lead and the trace. If you are using a probe of some type, the situation is worse; probes have a contact resistance that is often a function of the applied pressure. Perhaps you have seen this when you have measured resistance using an Ohmmeter.

The method shown in Figure A2.2 avoids this problem. In fact, the method shown in Figure A2.2 is very similar to the method the IPC uses in their thermal testing (reference Figure 4.5, the IPC standard test trace.)

A2.3: Problem with Ohmmeter Measurement

Figures A2.3 and A2.4 illustrate two common forms of Ohmmeter configurations, series and shunt [1]. In the series design, R1 and the adjustable resistor are selected so that when the resistor under test (Rx) is shorted the meter reads full scale. This is defined as the zero-resistance point on the scale. When Rx is infinite (i.e. open), no current flows. When Rx equals R1 plus the adjustable resistance, the meter reads half scale.

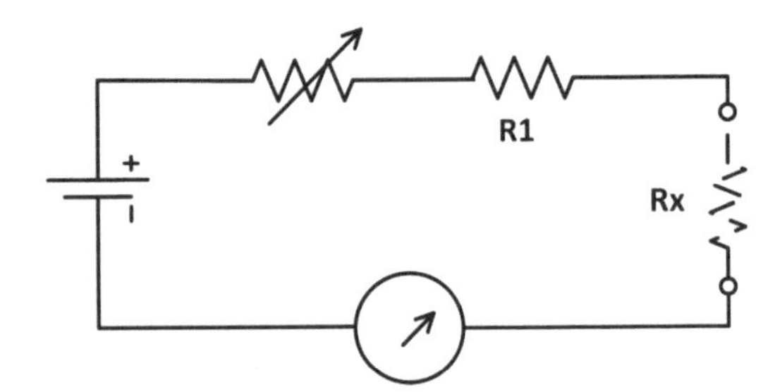

Figure A2.3
Representative series-type ohmmeter design

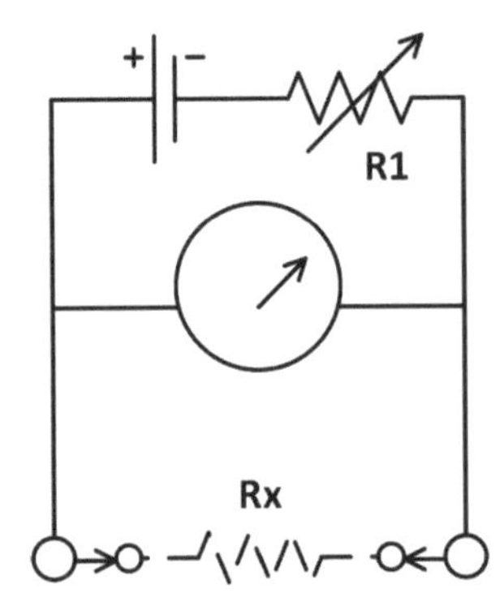

Figure A2.4
Representative shunt-type ohmmeter design

In the shunt-type configuration, when Rx is shorted, no current flows through the meter. When Rx is open, then R1 is adjusted so the meter reads full scale. When Rx equals R1, the meter reads one-half scale.

The problem with typical Ohmmeters is that the current through Rx is usually quite small (meter movements are typically 50 uA to 200 uA full scale), violating the principle we postulated above that the maximum practical parameters (in this case current) should be used.

There are four-terminal ohmmeters available. Some people believe that using a 4-terminal ohmmeter solves any ohmmeter problems with these types of measurements. The difference between two-terminal and four-terminal meters is illustrated in Figures A2.5 and A2.6. But note that the difference between the measurement techniques shown in Figures A2.5 and A2.6 is *exactly* the same as the difference in the measurement techniques shown in Figures A2.1 and A2.2! Figure A2.6 looks just like the technique of applying a known current through the unknown resistance and measuring the voltage across it. But the problem of small signal levels still remains [2].

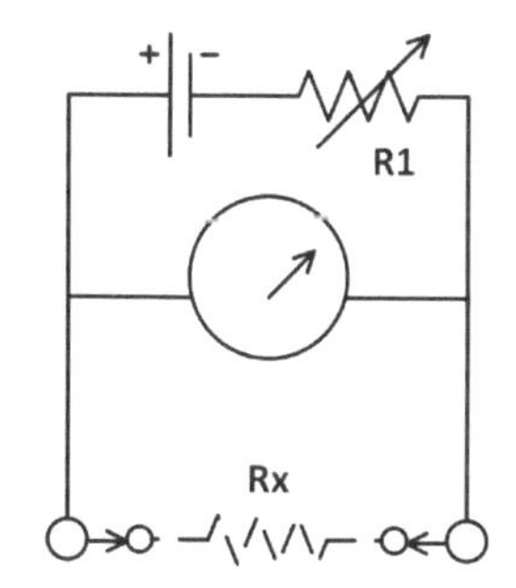

Figure A2.5
Typical 2-wire ohmmeter.

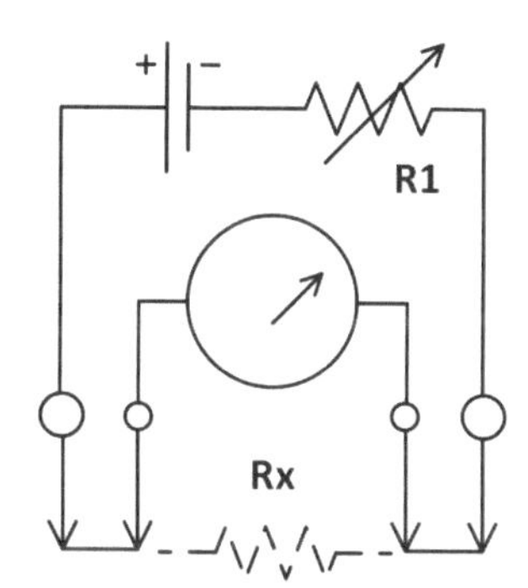

Figure A2.6
Typical 4-wire ohmmeter.

A2.4: Sources of Measurement Error

It is one thing to measure trace resistance. But then we need to calculate resistivity from there using Equation A2.1. In order to do that we need to know the width of the trace, the length of the trace, and the thickness of the trace. *We cannot take the nominal design dimensions for these values.* The three dimensions are discussed below.

Trace width: Trace width is probably the easiest dimension to deal with. We actually can use the design dimension here. The tolerance on trace width should be within 0.1 mil, which results in a measurement tolerance of about 1.0% in even the worst of cases.

Trace length: Length is a little trickier. Figure A2.2 shows where the length should be measured from, and if the trace looks like that shown in the figure, there should not be much of an issue. But if there is a pad at the end of the trace (see Figure A2.7), there can be a problem.

Figure A2.7
Trace and pad.

The pad can be either circular or rectangular. If the contact point (circle) is at the forward edge of the pad, the measurement will probably work out. I have found experimentally, that if the pad is *very* much wider than the trace, then the pad resistance is probably so low that there is no issue, either. But if there is a pad (similar to that shown) and the pad width is not very much wider than the trace, then there will be current densities that will be very difficult to adjust for. The resistance of the pad will impact the measurement.

Trace thickness: But by far the most difficult issue to deal with is trace thickness. If there is copper plating on the trace, trace thickness can vary by as much as 50% at various places around the board. A 50% on trace thickness results in a 50% tolerance on the cross-sectional area calculation, which will result on a 50% tolerance on the resistivity calculation. Even copper foil typically has a 10% tolerance.

One way to deal with this is to microsection the board where the measurements are being made. But microsections are only valid at the point where they are taken. We can only hope they are representative of other places along the trace. I will show you in the next section that trace thickness even varies across the trace at any given point!

Roughness: Figure A2.8 illustrates the roughness between the copper foil and laminate for a 2.0 Ounce trace. It is my experience, through measurement and simulations, that roughness does not have a significant impact on DC trace resistance [3].

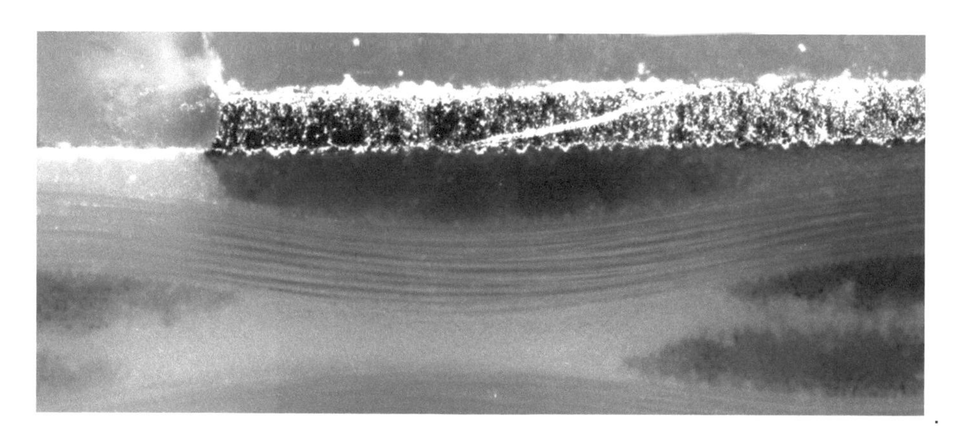

Figure A2.8
Roughness between a 2.0 Oz foil and the board laminate.

A2.5: An Experimental Study

Prototron Circuits was generous enough to donate a board (and microsectioning services) for evaluating trace resistivity. The design criteria incorporated the idea that all dimensions should be as large as practical. The design criteria was as follows:

Width: Traces were 200 mil wide.
Length: Traces were designed to look like Figure A2.2
Trace length was 5.0"
Thickness: Three thicknesses were employed.
Copper foil (2.0 Oz, nominally 2.3 mils thick)
Copper foil + 1.0 Oz plated copper (total nominally 3.9 mils thick)
Copper foil + 1.0 Oz plated copper, + 1.0 Oz additional plated copper (total nominally 5.2 mils thick.)

Traces were grouped in sets of three (the three thicknesses) and repeated around the board. Five sets of traces were analyzed, four on the top layer and one on the bottom layer.

Each set of traces was microsectioned. Figure A2.9 is a microscopic view of the edge of one of the three-layer, 4 Oz., traces. The lines between the three trace layers are visible in this view. It is also evident that the trace is thicker at the edge than it is a few mils in from the edge, illustrating the fact that the trace thickness can vary even from edge to edge. Figure A2.10 is a microscopic view of a two-layer, 3Oz., trace. The difference in thickness between the edge of the trace and near the midpoint is also evident in that trace. In general, foil traces were pretty flat. Traces with copper plating applied often had some variation in thickness from edge to edge, as much as 10% in some cases.

Figure A2.9
Three layer trace,

Figure A2.10
Two-layer trace.

The thickness of each microsectioned trace was measured at 6 place across the width. The extremes of the measurements (across all traces) are shown in Table A2.1. The measurement tolerance is approximately 0.2 to 0.3 mils. As can be seen there is a very wide variation in thickness for traces that are nominally the same.

	2 Oz.	3 Oz.	4 Oz.
Max	2.8	4.3	5.7
Min	1.9	3.0	4.3
Overall average	2.4	3.6	4.7

Table A2.1
Thickness variation of traces (in mils)

A current of 2.1 Amps was applied to the trace and the voltage across the trace was measured. This level of current was sufficiently high to allow reasonably precise voltage measurements while not high enough to heat the traces. That is, the traces remained at the ambient room temperature. The resistance was calculated from current divided by voltage (Ohm's Law), and the resistivity calculated (based on the nominal width and length, and the average measured thickness of each trace.) The resistivities were averaged for each thickness. The resistivities were then adjusted to 20° C (the room temperature was 24.5° C when the data were taken) using Equation 3.6. The actual resistivities were compared to the expected resistivities based on the hypothesis that the resistivity of the plated copper was that of pure copper, 1.7 uOhm-cm. Table A2.2 summarizes the results.

	2 Oz.	3 Oz.	4 Oz.
Average resistivity	1.806	1.786	1.766
Adjusted to 20° C	1.788	1.758	1.739
"Expected resistivity"	1.788 *	1.752	1.739
* taken as the expected value for the 2 Oz. foil			

Table A2.2

Calculated resistivities, uOhm-cm.

What is "expected resistivity"? Think of the three layers of copper as parallel resistors. The first layer is the copper foil. It has an adjusted resistivity of 1.788 uOhm-cm. Let's assume the second layer is pure copper. It would have a resistivity of 1.7 uOhm-cm. In terms of parallel resistors, we have one resistor (2 Oz. thick) of one resistivity, and a second parallel resistor of resistivity 1.7 uOhm-cm. They would combine to have an expected resistivity as shown in the table. We compare that to the measured resistivity ("adjusted to 20° C") and would expect the values to be the same if our assumed value for the resistivity of plated copper is correct. As can be seen the values agree very closely.

A2.6: Internal Traces

There are a couple of references on the web where the authors suggest that internal traces exhibit higher resistivity than do external traces. The reasons have to do with fabrication processes (although the specifics are unstated.)

Another board was fabricated to evaluate this claim. The internal traces were fabricated just like the external traces evaluated above with the following exceptions:

1. Because of space limitations, the traces were only 100 mils wide.
2. Because of fabrication limitations associated with the inner layers, only two thicknesses could be evaluated.

The foil layers were measured as 1.5 Oz., or 2.0 mils. There was some variability among the internal traces as to trace thickness, and there may be an increased measurement tolerance because of that fact. The average thickness of the plated traces was 4.2375 mils. Subtracting the foil thickness leaves the average plating thickness as 2.2375 mils.

A current of 1.5 A. was applied to the internal traces and the resistivity calculated using the same procedures as above. The results are shown in Table A2.3.

	Foil	Plated trace
Average resistivity uOhm-in	0.719	0.709
Adjusted to 20° C	0.706	0.695

Table A2.3
Average resistivities of internal traces

From these results we can calculate the resistivity of the plated copper layer over the foil. That calculation results in a value of 0.686 uOhm-in

A2.7: Summary

Table A2.4 summarizes the resistivity calculations for the copper foil and for the copper plating.

There is no evidence from this experimental investigation that the resistivity of PCB traces is anything other than what would be expected. In particular, the plated copper appears to have a resistivity of around 0.68 uOhm-in or 1.7 uOhm-cm, that of pure copper (given the measurement tolerances involved.)

	Foil		Plated	
	uOhm-in	uOhm-cm	uOhm-in	uOhm-cm
External	0.704	1.788	0.670	1.70
Internal	0.706	1.790	0.695	1.765
"Published" *	0.68-0.70	1.72-1.77	0.66	1.68

Table A2.4
Calculated values of resistivity (* = generally accepted values)

Notes:

1. See https://www.edgefx.in/ohmmeter-circuit-and-types-of-ohmmeters/ for a description of series and shunt ohmmeters.
2. There are special purpose four-terminal meters available that can supply relatively large currents (2.0 or 4.0 Amps) to the resistance under test.
3. It is known, of course, that roughness has an impact on high-speed traces and designs. But our context here is DC.

A3 IPC CURVES

INTERNAL and VACUUM IPC CURVES FITTED WITH EQUATIONS

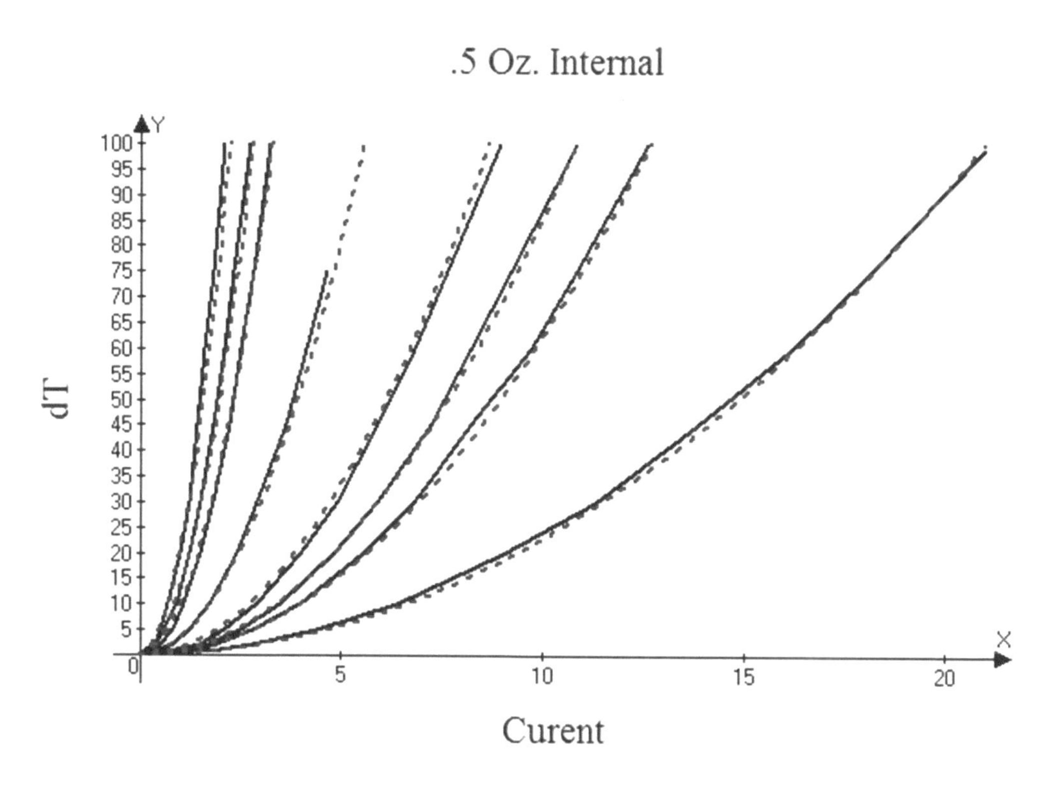

0.5 Oz. internal data fitted with equations.

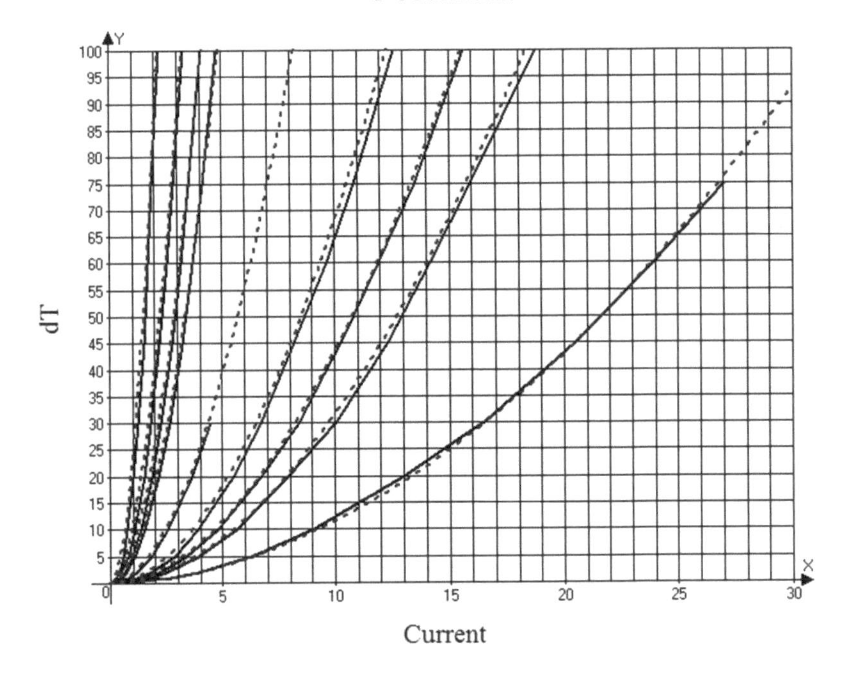

1 Oz. Internal data fitted with equations.

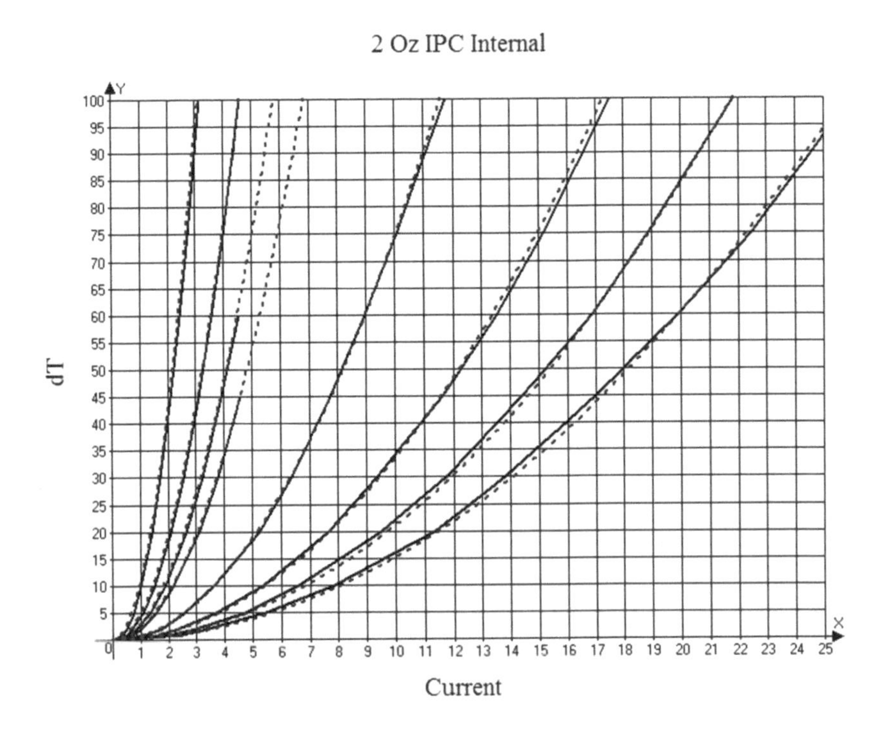

2 Oz. internal data fitted with equations

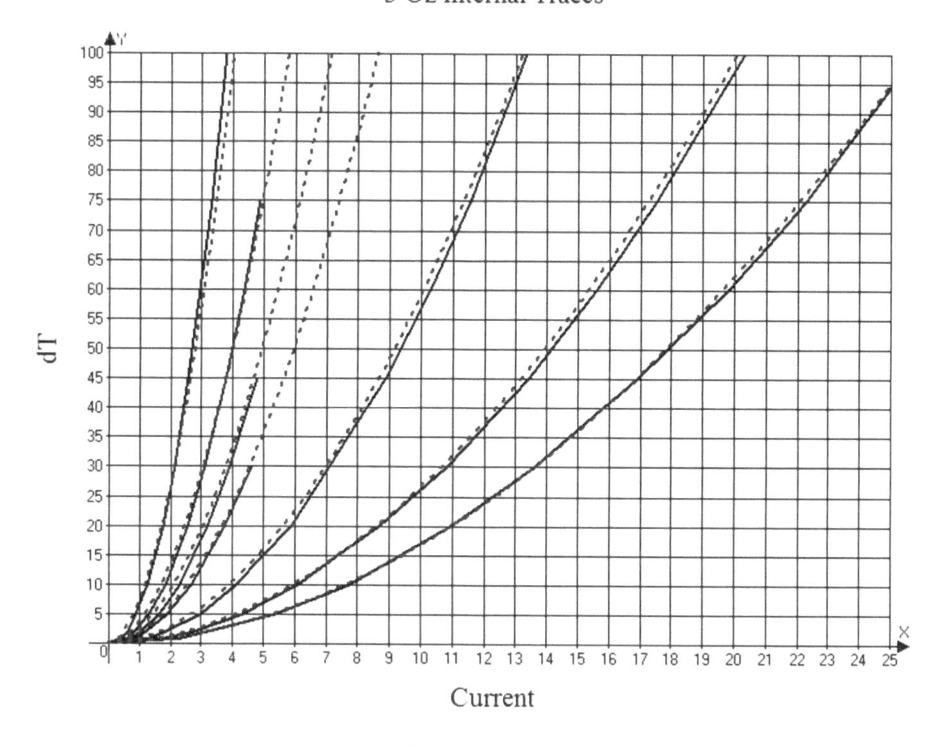

3 Oz. internal data fitted with equations.

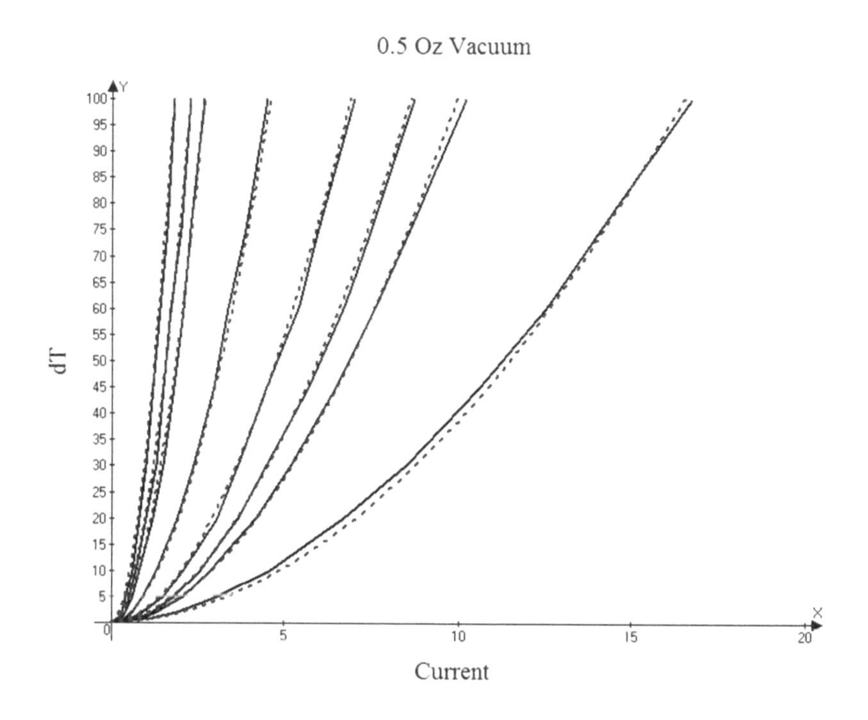

0.5 Oz. vacuum data fitted with equations.

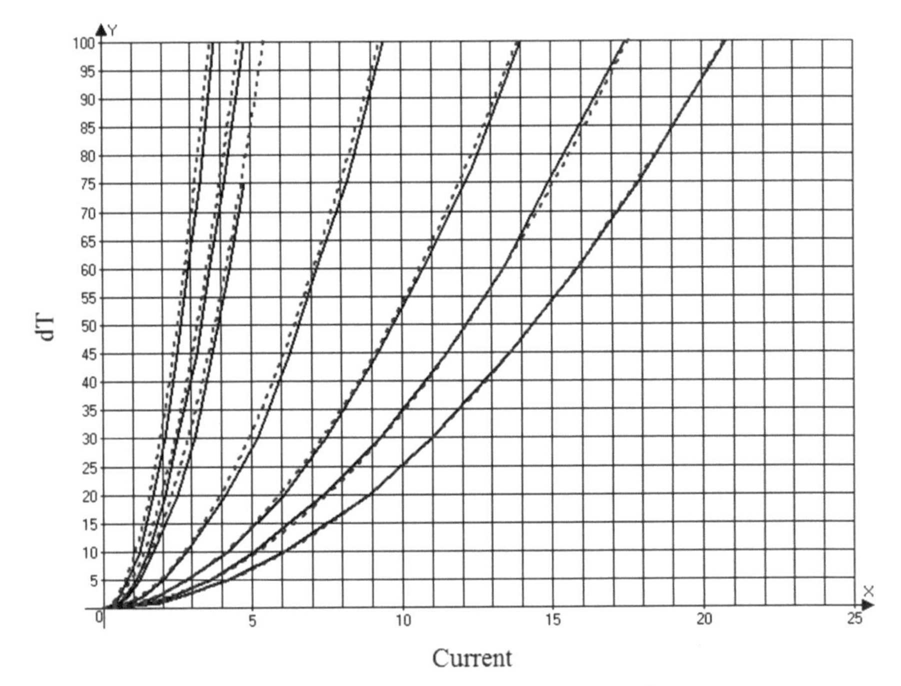

2 Oz. vacuum data fitted with equations.

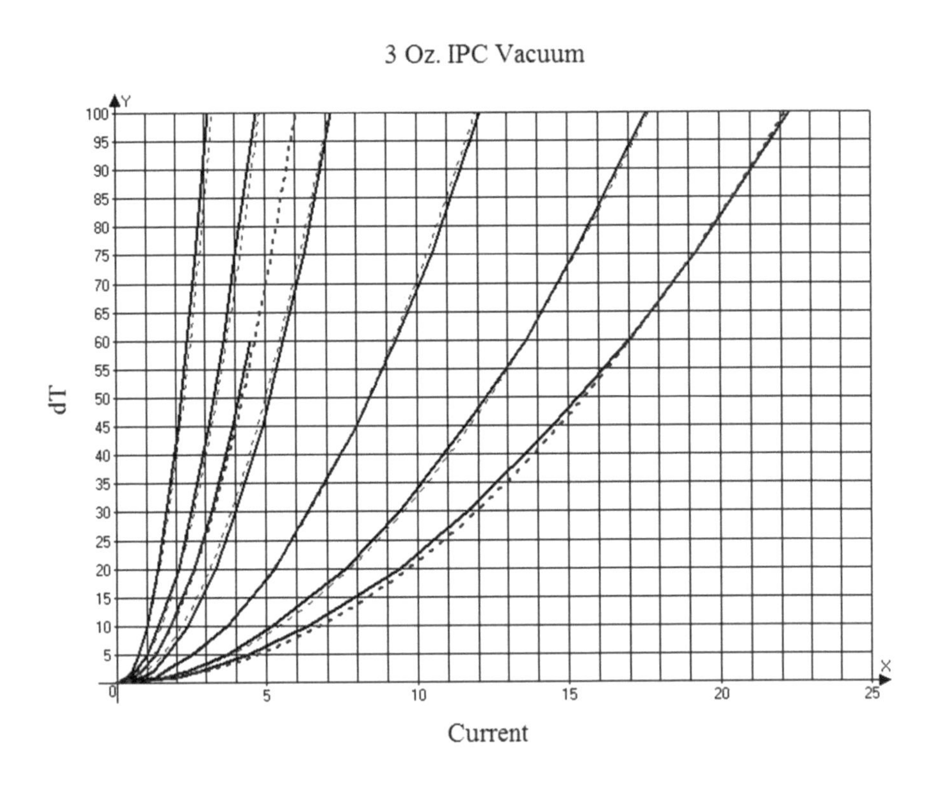

3 Oz. vacuum data fitted with equations.

A4 EQUATIONS

DETAILED SETS OF EQUATIONS FOR THE CURVES

In the few cases where there are differences by width within a trace thickness, the differences are small and probably reflect errors, uncertainties, and variations as a result of various graphical drawings and manipulations.

Structure **External**	**dT = Constant**	**W^**	**Th^**	**C^**	
All Thicknesses	215.3	-1.15	-1.00	2	
Internal					
.5 Oz	110	-1.10	-1.52	2	for 100 mil and wider
	125	-1.10	-1.52	2	50 mil
	130	-1.10	1.52	2	20 mil and smaller
1 Oz	200	-1.10	-1.52	1.9	
2 Oz	300.3	-1.15	-1.52	2	
3 Oz	300	-1.15	-1.52	1.9	for 50, 100, 150 mil
	200	-1.15	-1.52	1.9	5 Mil
	225	-1.15	-1.52	1.9	10 Mil
	240	-1.15	-1.52	1.9	15 Mil
	235	-1.15	-1.52	1.9	20 Mil
Vacuum					
.5 Oz.	210	-1.10	-1.52	1.9	100 mil and smaller
	215	-1.10	-1.52	1.9	150 mil
	225	-1.10	-1.52	1.9	200 mil
	235	-1.10	-1.52	1.9	500 mil
2 Oz.	480	-1.10	-1.52	1.9	
3 Oz	460	-1.10	-1.52	1.95	

A5 INTERNAL and VACUUM IPC CURVES

Internal 1 Oz., 2 Oz., and 3 Oz. curves:

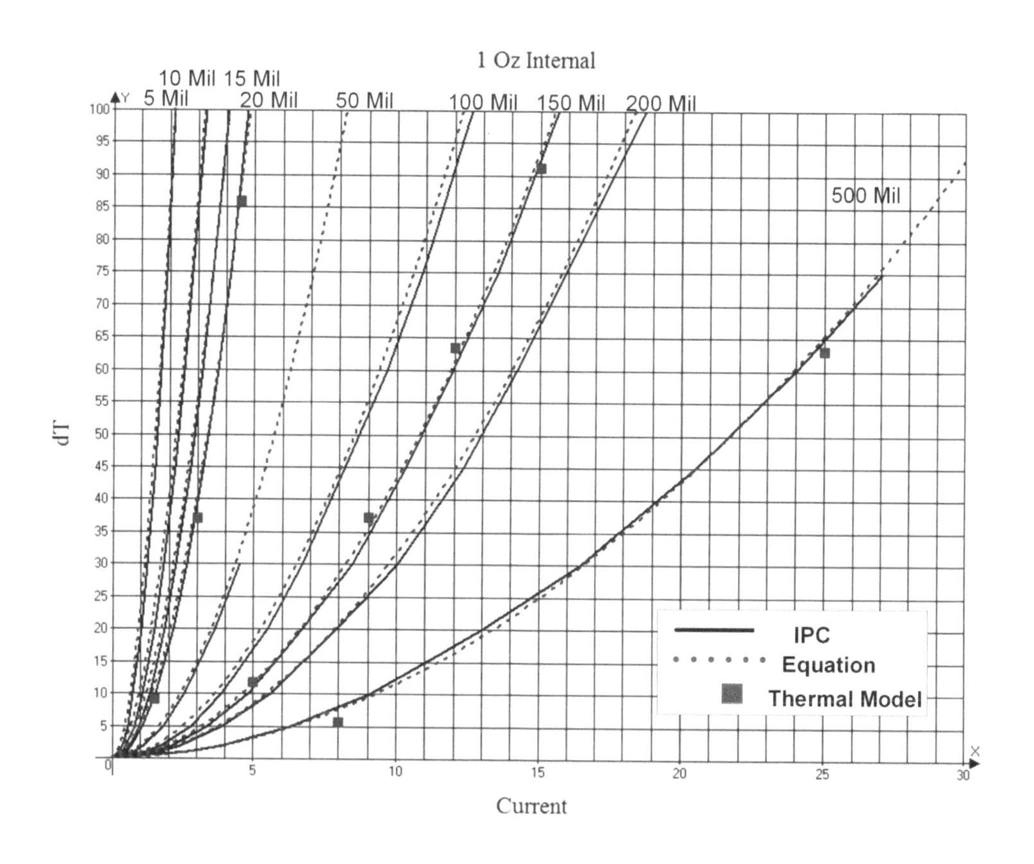

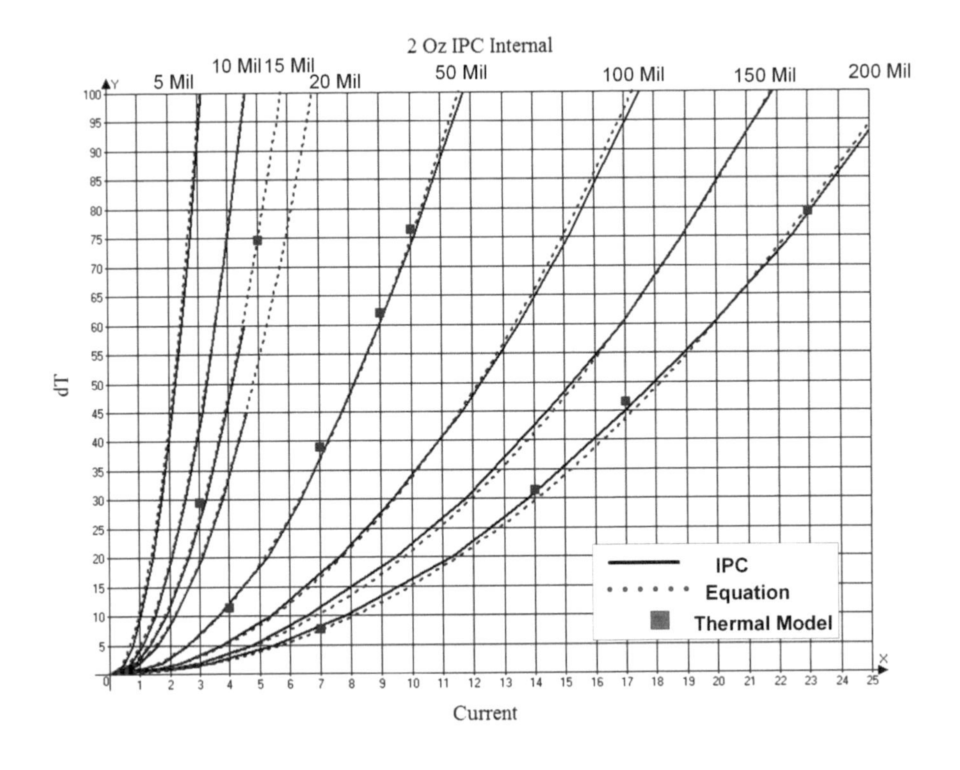
2 Oz IPC Internal
5 Mil
10 Mil
15 Mil
20 Mil
50 Mil
100 Mil
150 Mil
200 Mil
dT
Current
IPC
Equation
Thermal Model

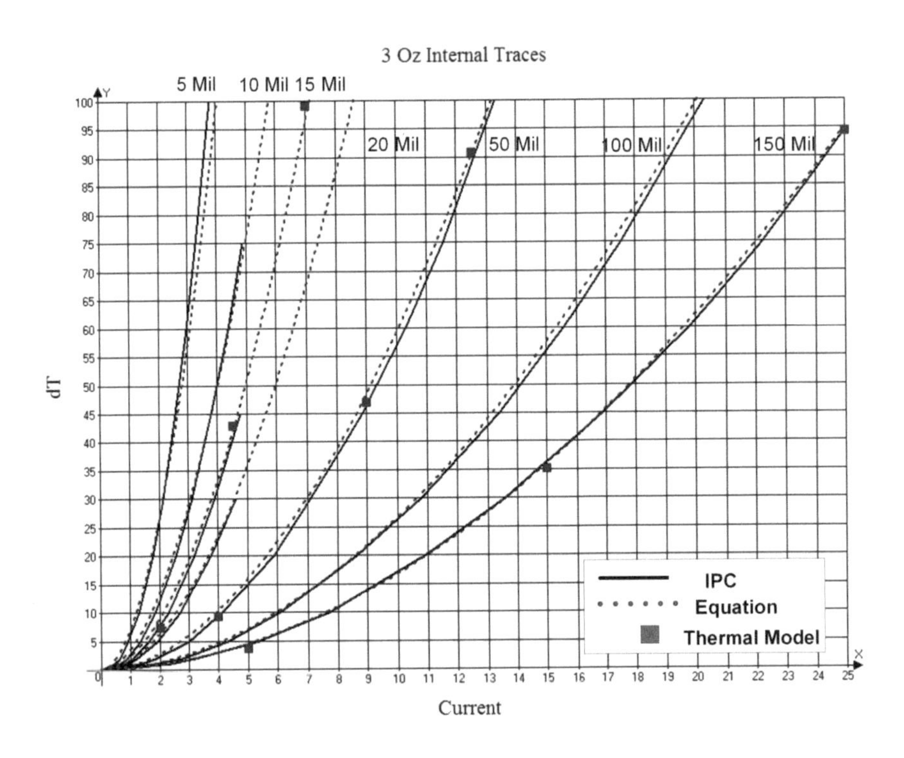
3 Oz Internal Traces
5 Mil
10 Mil
15 Mil
20 Mil
50 Mil
100 Mil
150 Mil
dT
Current
IPC
Equation
Thermal Model

Vacuum 2 Oz. curves:

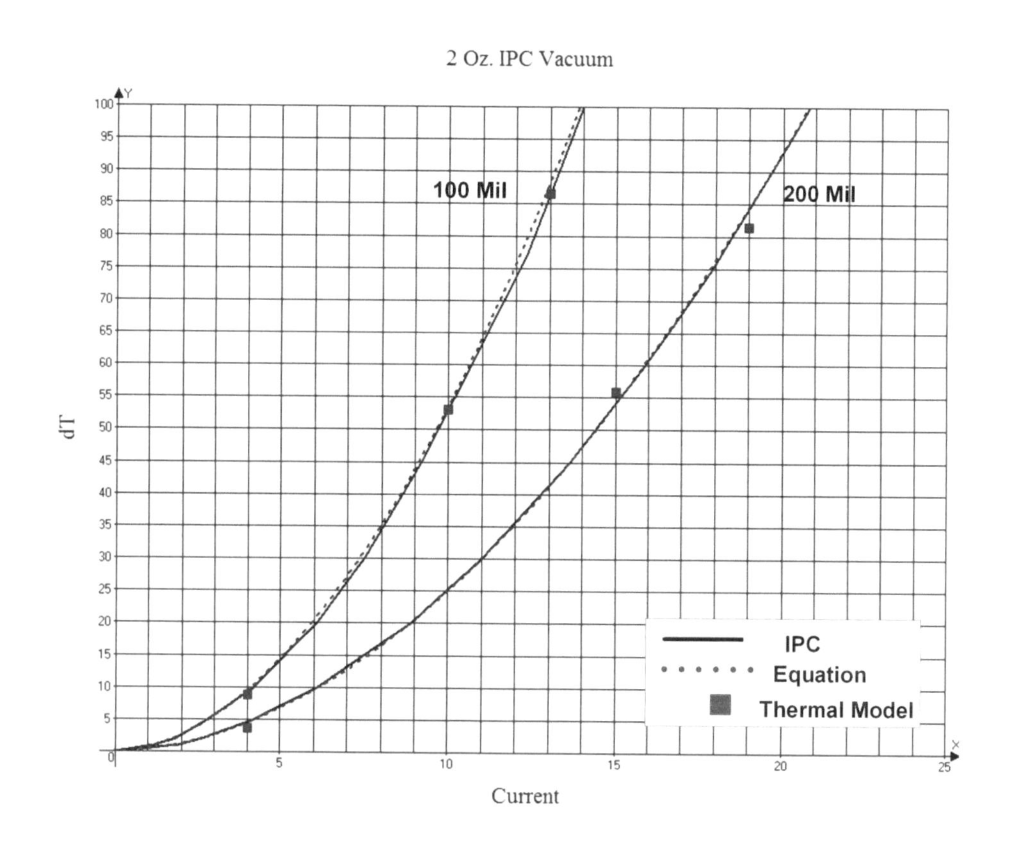

A6 CURRENT DENSITY IN VIAS

Appendix A6.1: How to interpret the via density patterns and analyses.

The following appendices look at the current densities (A/mm^2) at various layers in the via. Typically, the “Top” layer is the trace layer and Core 1 through

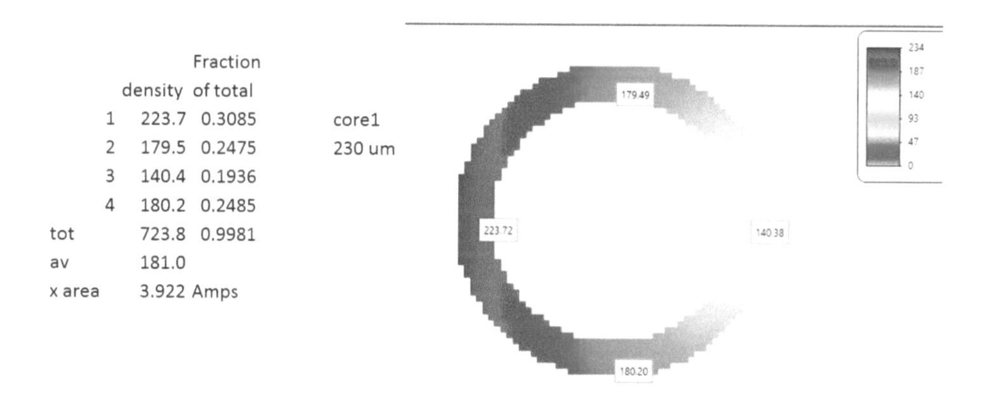

		density	Fraction of total
	1	223.7	0.3085
	2	179.5	0.2475
	3	140.4	0.1936
	4	180.2	0.2485
tot		723.8	0.9981
av		181.0	
x area		3.922	Amps

Core 4 are the first four (of seven) dielectric layer models. The following figure will illustrate the analyses that are going on at each layer for each simulation. This figure is for the first core layer in the single via model.

At each layer, four current density readings are taken around each layer. They are labeled 1 through 4 in the column in the table. Reading 1 corresponds to the left side of the via; 2 corresponds to the top of the via; 3 corresponds to the right-hand side of the via; and 4 corresponds to the bottom of the via. The density readings are recorded in the table and summed in the “tot” row. The column on the right is the fraction that that density is of the density “tot.” We would expect the fraction total to equal 1.0. The average density (“av” in the table) is the total of the density readings divided by 4 (the number of readings.) The “x area” is the product of the average density times the via conducting area, which is always .0217 mm^2. This should equal the total current through the via (in this case 4.0 Amps.)

CAUTION: A major word of caution is required here. Current density is a "point" concept. The density has a value at a specific point, and may vary from point-to-point. In order to obtain the total current in the via wall at any layer we must *integrate* the density around the circular wall of the via. What we are doing, taking only four readings, is an approximation. We believe four readings are reasonable because the variation in the density is smooth and continuous. But it is expected that our results will be approximate.

The current density patterns are symmetrical, so we need only look at the first 4 core layers. The remaining layers are identical in reverse order. In the case of 4 vias in the straight line configuration, the density patterns are symmetrical through Core 4 and also symmetrical around the center-line of the trace. In the case of 4 vias turning at an angle, there is a symmetry if the viewer looks at it as "twisting" as we go down through the via layers. So in all cases we need only look at the core layers down through the middle one. The following illustrates the symmetry between the top (left) and bottom (right) layers in the single via model.

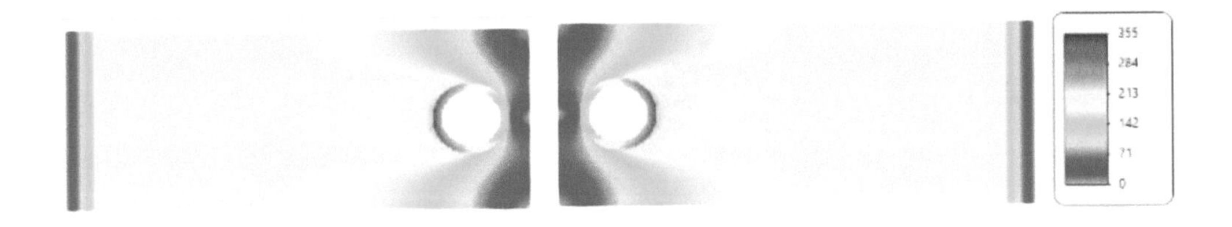

Appendix A6.2

Single VIA Model

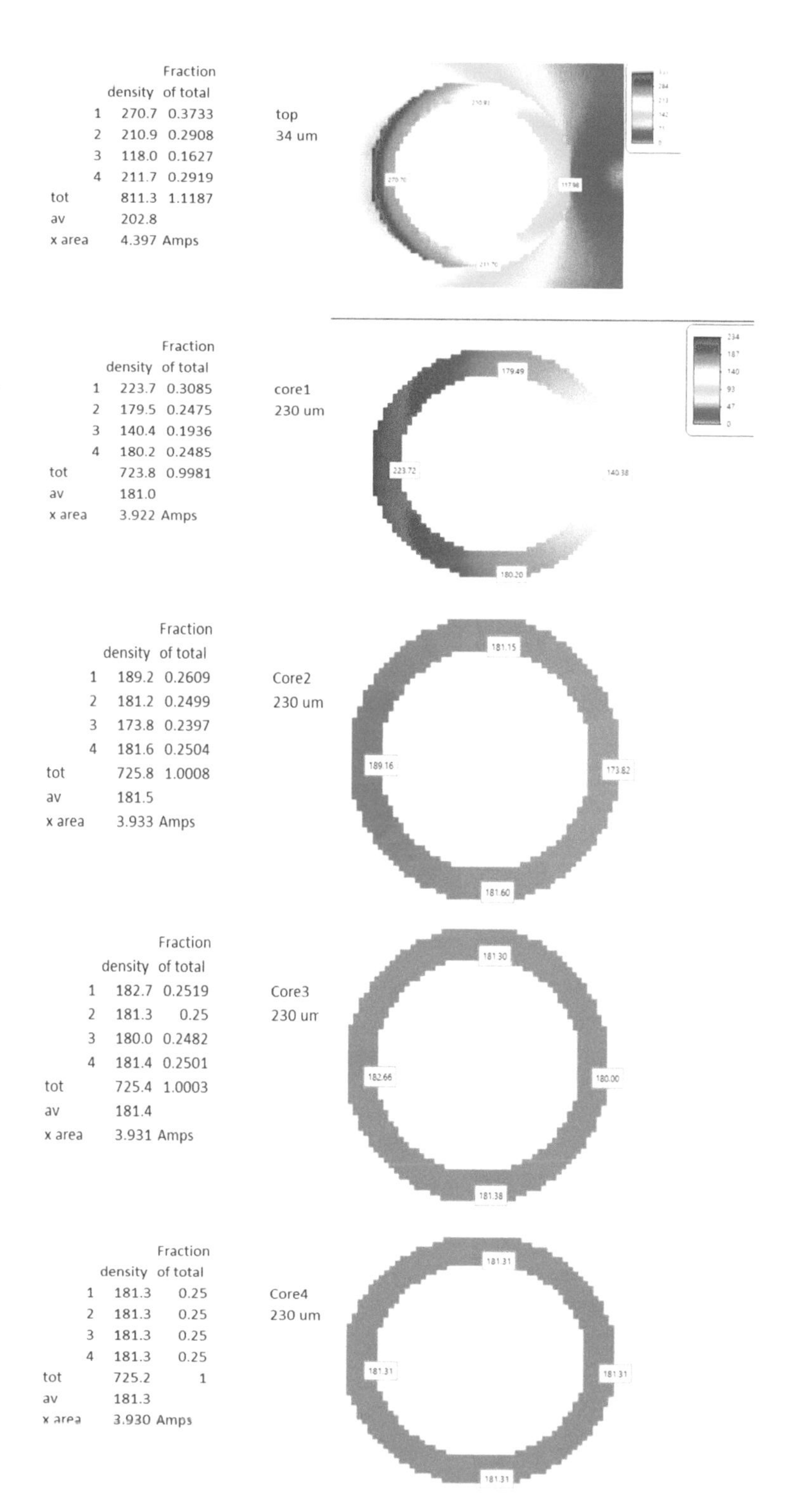

top 34 um

	density	Fraction of total
1	270.7	0.3733
2	210.9	0.2908
3	118.0	0.1627
4	211.7	0.2919
tot	811.3	1.1187
av	202.8	
x area	4.397	Amps

core1 230 um

	density	Fraction of total
1	223.7	0.3085
2	179.5	0.2475
3	140.4	0.1936
4	180.2	0.2485
tot	723.8	0.9981
av	181.0	
x area	3.922	Amps

Core2 230 um

	density	Fraction of total
1	189.2	0.2609
2	181.2	0.2499
3	173.8	0.2397
4	181.6	0.2504
tot	725.8	1.0008
av	181.5	
x area	3.933	Amps

Core3 230 um

	density	Fraction of total
1	182.7	0.2519
2	181.3	0.25
3	180.0	0.2482
4	181.4	0.2501
tot	725.4	1.0003
av	181.4	
x area	3.931	Amps

Core4 230 um

	density	Fraction of total
1	181.3	0.25
2	181.3	0.25
3	181.3	0.25
4	181.3	0.25
tot	725.2	1
av	181.3	
x area	3.930	Amps

Appendix A6.3

Single via model with Core 1 broken into three cores, the first two with 15 um thicknesses.

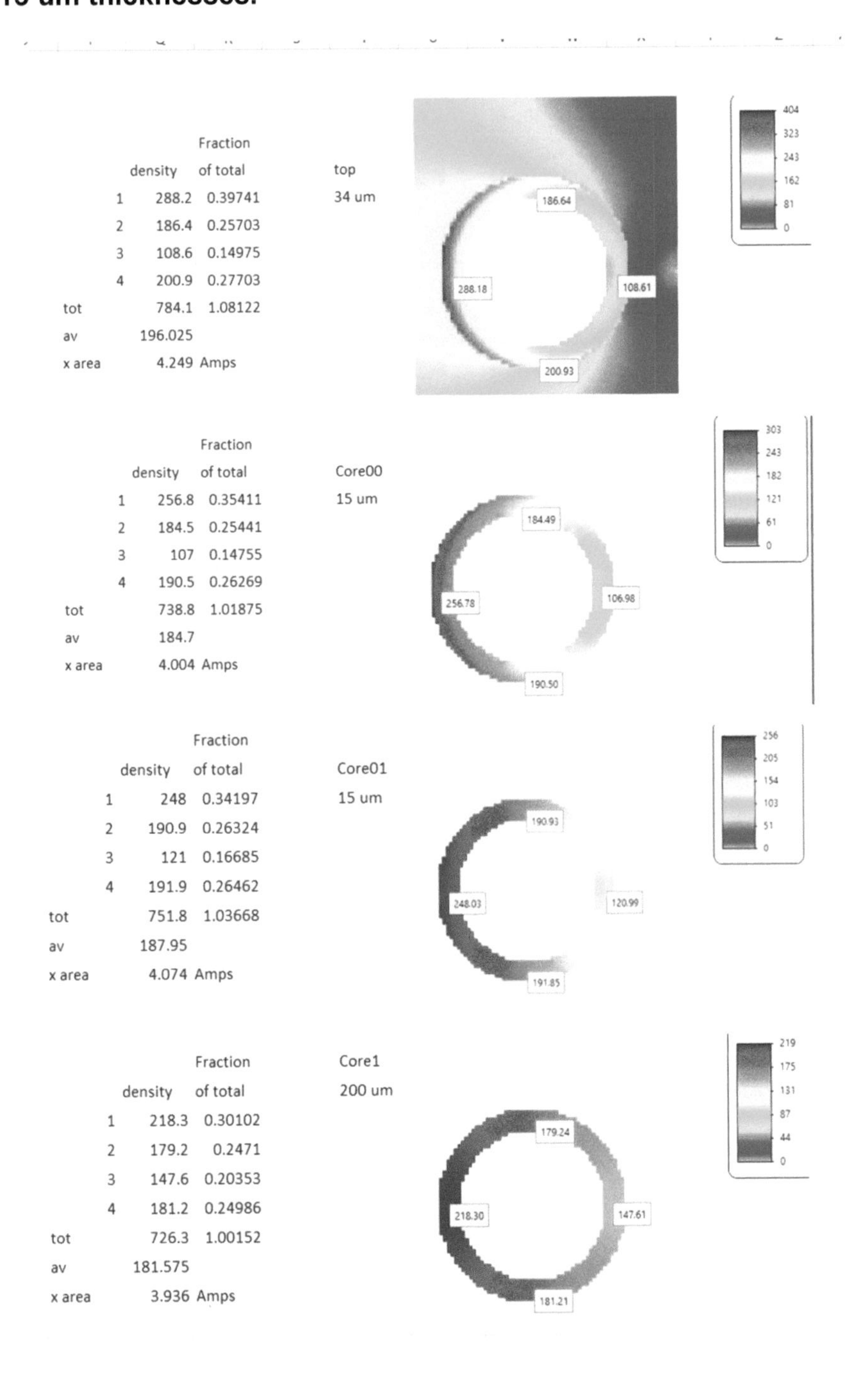

top 34 um

	density	Fraction of total
1	288.2	0.39741
2	186.4	0.25703
3	108.6	0.14975
4	200.9	0.27703
tot	784.1	1.08122
av	196.025	
x area	4.249	Amps

Core00 15 um

	density	Fraction of total
1	256.8	0.35411
2	184.5	0.25441
3	107	0.14755
4	190.5	0.26269
tot	738.8	1.01875
av	184.7	
x area	4.004	Amps

Core01 15 um

	density	Fraction of total
1	248	0.34197
2	190.9	0.26324
3	121	0.16685
4	191.9	0.26462
tot	751.8	1.03668
av	187.95	
x area	4.074	Amps

Core1 200 um

	density	Fraction of total
1	218.3	0.30102
2	179.2	0.2471
3	147.6	0.20353
4	181.2	0.24986
tot	726.3	1.00152
av	181.575	
x area	3.936	Amps

Note that the current density effects unique to the top (trace) layer continue into the first small layer, and to a lesser extent into the second small layer. After that, the current densities are essentially the same as with the simulation shown in Appendix A6.2.

Appendix 6.4

Simulation of four vias, proceeding straight ahead.

The patterns are symmetrical around the horizontal center line of the trace.

		density	Fraction of total
	1	136.87	0.753689
	2	129.2	0.711454
	3	33.77	0.185958
	4	124	0.682819
tot		423.84	2.333921
av		105.96	
x area		2.299332	Amps

		density	Fraction of total
	1	76.4	0.420705
	2	45.1	0.248348
	3	29.6	0.162996
	4	66.8	0.367841
tot		217.9	1.19989
av		54.475	
x area		1.182108	Amps

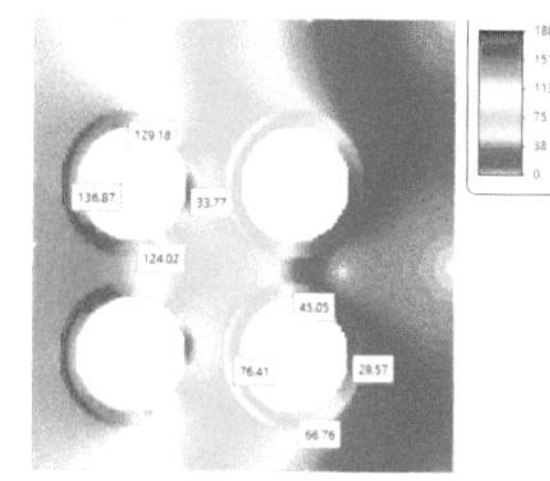

Trace Layer

		density	Fraction of total
	1	95.5	0.525881
	2	58.3	0.321035
	3	2.01	0.011068
	4	48	0.264317
tot		203.81	1.122302
av		50.9525	
x area		1.105669	Amps

		density	Fraction of total
	1	60.8	0.334802
	2	37.1	0.204295
	3	34.2	0.188326
	4	49.4	0.272026
tot		181.5	0.999449
av		45.375	
x area		0.984638	Amps

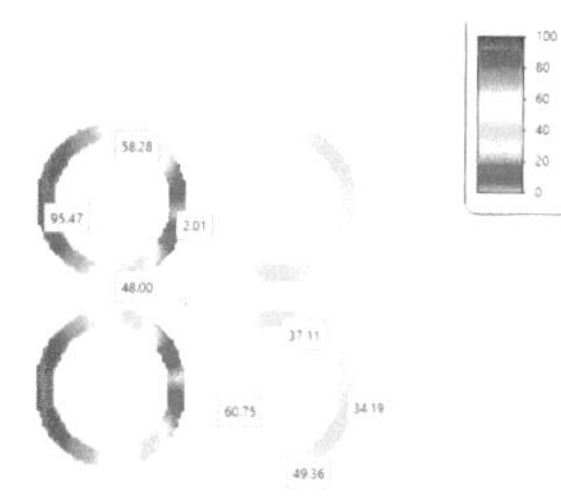

Core 1

		density	Fraction of total
	1	53.7	0.295705
	2	46.6	0.256608
	3	36.9	0.203194
	4	44.5	0.245044
tot		181.7	1.000551
av		45.425	
x area		0.985723	Amps

		density	Fraction of total
	1	48	0.264317
	2	44	0.242291
	3	43.1	0.237335
	4	46.5	0.256057
tot		181.6	1
av		45.4	
x area		0.98518	Amps

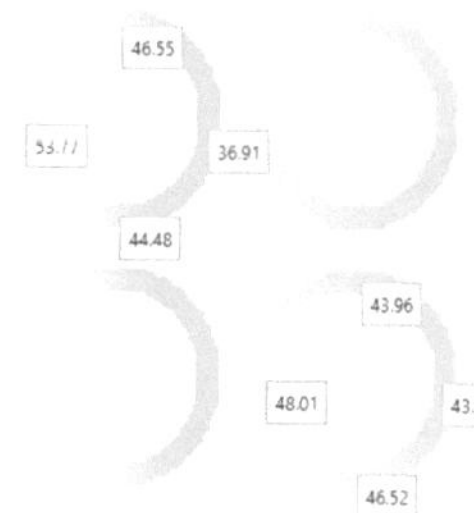

Core 2

	density	Fraction of total
1	46.8	0.257709
2	45.5	0.250551
3	43.8	0.241189
4	45.1	0.248348
tot	181.2	0.997797
av	45.3	
x area	0.98301	Amps

	density	Fraction of total
1	45.8	0.252203
2	45.2	0.248899
3	45	0.247797
4	45.6	0.251101
tot	181.6	1
av	45.4	
x area	0.98518	Amps

Core 3

	density	Fraction of total
1	45.5	0.250551
2	45.4	0.25
3	45.1	0.248348
4	45.2	0.248899
tot	181.2	0.997797
av	45.3	
x area	0.98301	Amps
total current		3.93638

	density	Fraction of total
1	45.2	0.248899
2	45.3	0.249449
3	45.6	0.251101
4	45.5	0.250551
tot	181.6	1
av	45.4	
x area	0.98518	Amps

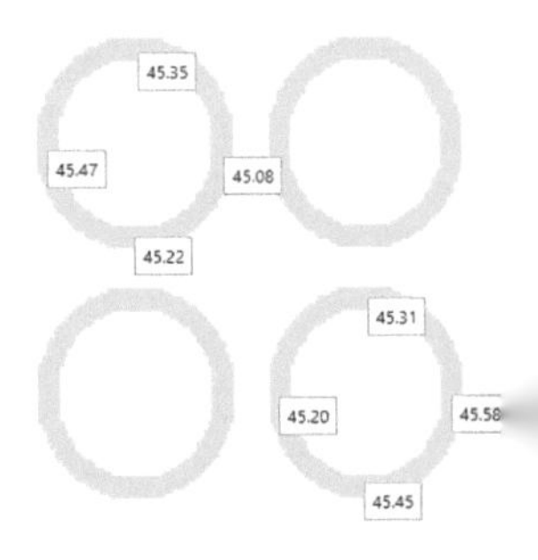

Core 4

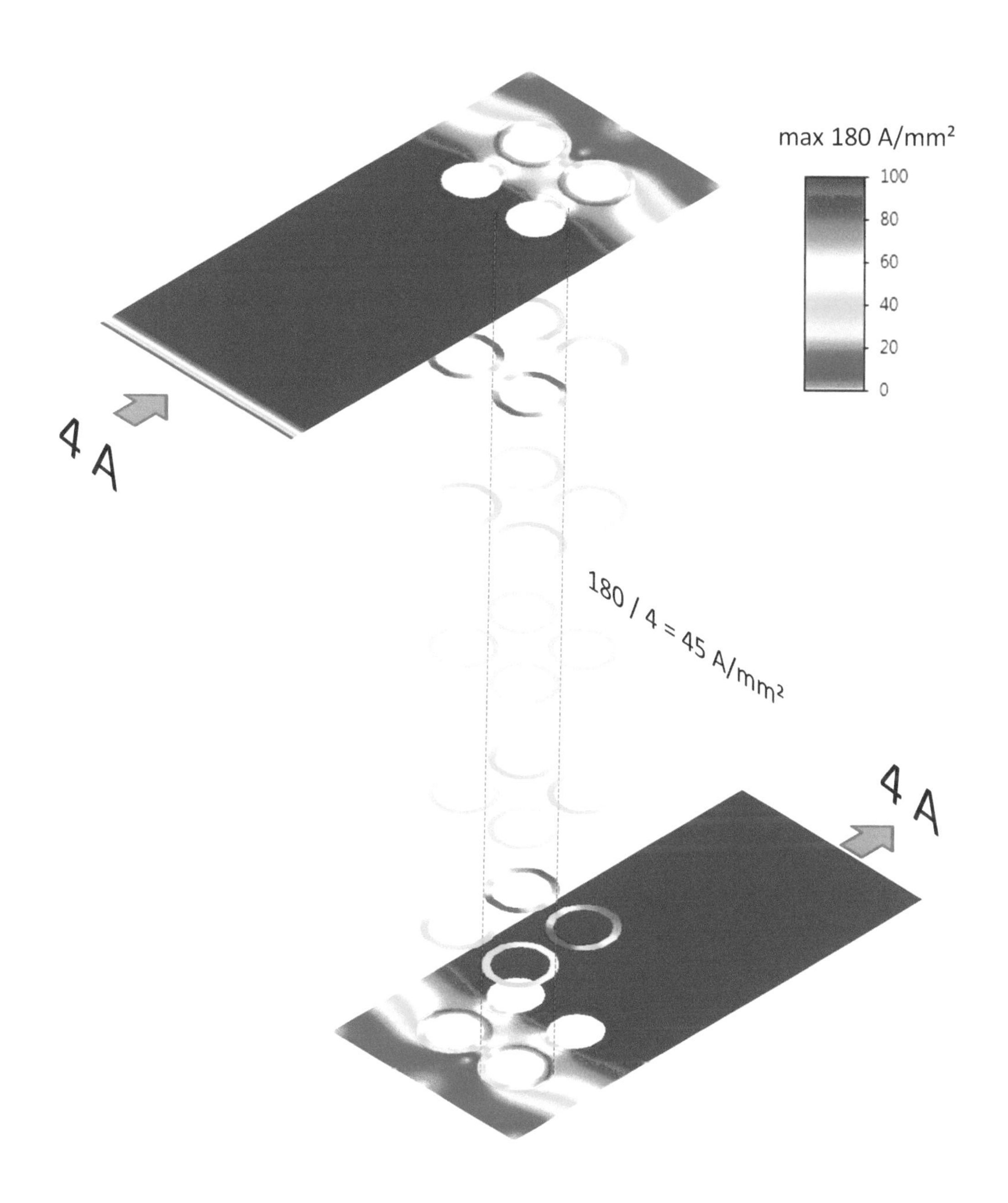

3D Representation of the layers

Courtesy Dr. Johannes Adam

Appendix A6.5

Simulation of four vias, traces at right angles.

There is a rotational symmetry in this model as the model turns 90 degrees from top layer to bottom layer.

Trace Layer

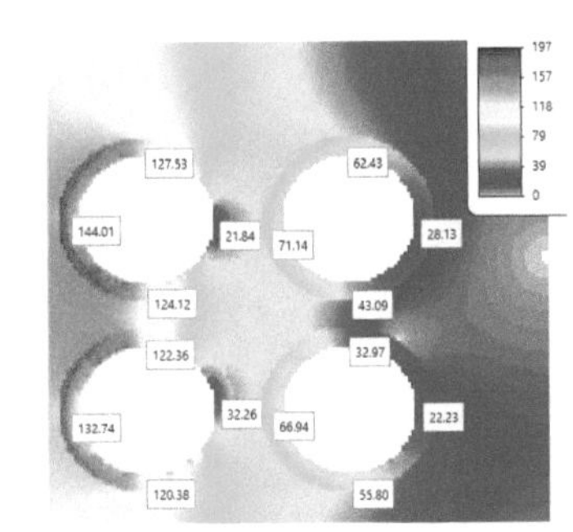

Top left

	density	fraction of total
1	144	0.792952
2	127.5	0.702093
3	21.8	0.120044
4	124.1	0.68337
tot	417.4	2.298458
av	104.35	
x area	2.264395	Amps

Top right

	density	fraction of total
1	71.1	0.39152
2	62.4	0.343612
3	28.1	0.154736
4	43.1	0.237335
tot	204.7	1.127203
av	51.175	
x area	1.110498	Amps

bottom left

	density	fraction of total
1	132.7	0.730727
2	122.4	0.674009
3	32.3	0.177863
4	120.4	0.662996
tot	407.8	2.245595
av	101.95	
x area	2.212315	Amps

bottom right

	density	fraction of total
1	66.9	0.368392
2	33	0.181718
3	22.2	0.122247
4	55.8	0.307269
tot	177.9	0.979626
av	44.475	
x area	0.965108	Amps

total current 6.552315

Core 1

Top left

	density	fraction of total
1	102.1	0.562225
2	66.5	0.366189
3	14.1	0.077643
4	55.7	0.306718
tot	238.4	1.312775
av	59.6	
x area	1.29332	Amps

Top right

	density	fraction of total
1	55.4	0.305066
2	47.5	0.261564
3	34	0.187225
4	38.4	0.211454
tot	175.3	0.965308
av	43.825	
x area	0.951003	Amps

bottom left

	density	fraction of total
1	89.6	0.493392
2	45.4	0.25
3	7.5	0.0413
4	57.7	0.317731
tot	200.2	1.102423
av	50.05	
x area	1.086085	Amps

bottom right

	density	fraction of total
1	47.1	0.259361
2	24.5	0.134912
3	24.1	0.132709
4	39.6	0.218062
tot	135.3	0.745044
av	33.825	
x area	0.734003	Amps

total current 4.06441

Appendix A6.5 (Continued)

Core 2

Top left

	density	fraction of total	
1	63.4	0.349119	
2	56	0.30837	
3	47.1	0.259361	
4	54.4	0.299559	
tot	220.9	1.21641	
av	55.225		
x area	1.198383	Amps	

Top right

	density	fraction of total
1	46.1	0.253855
2	44.8	0.246696
3	42	0.231278
4	43.2	0.237885
tot	176.1	0.969714
av	44.025	
x area	0.955343	Amps

bottom left

	density	fraction of total
1	52.5	0.289097
2	43.1	0.237335
3	36.7	0.202093
4	45.9	0.252753
tot	178.2	0.981278
av	44.55	
x area	0.966735	Amps

bottom right

	density	fraction of total
1	35.8	0.197137
2	31.9	0.175661
3	31.3	0.172357
4	34.8	0.19163
tot	133.8	0.736784
av	33.45	
x area	0.725865	Amps

total current 3.846325

Core 3

Top left

	density	fraction of total
1	56.4	0.310573
2	55.2	0.303965
3	53.5	0.294604
4	44.3	0.243943
t	209.4	1.153084
	52.35	
area	1.135995	Amps

Top right

	density	fraction of total
1	44.3	0.243943
2	44.1	0.242841
3	43.5	0.239537
4	43.7	0.240639
tot	175.6	0.96696
av	43.9	
x area	0.95263	Amps

bottom left

	density	fraction of total
1	45.6	0.251101
2	43.8	0.241189
3	42.7	0.235132
4	44.3	0.243943
tot	176.4	0.971366
av	44.1	
x area	0.95697	Amps

bottom right

	density	fraction of total
1	33.7	0.185573
2	33	0.181718
3	32.9	0.181167
4	33.5	0.184471
tot	133.1	0.73293
av	33.275	
x area	0.722068	Amps

tal current 3.767663

Appendix A6.5 (Continued)

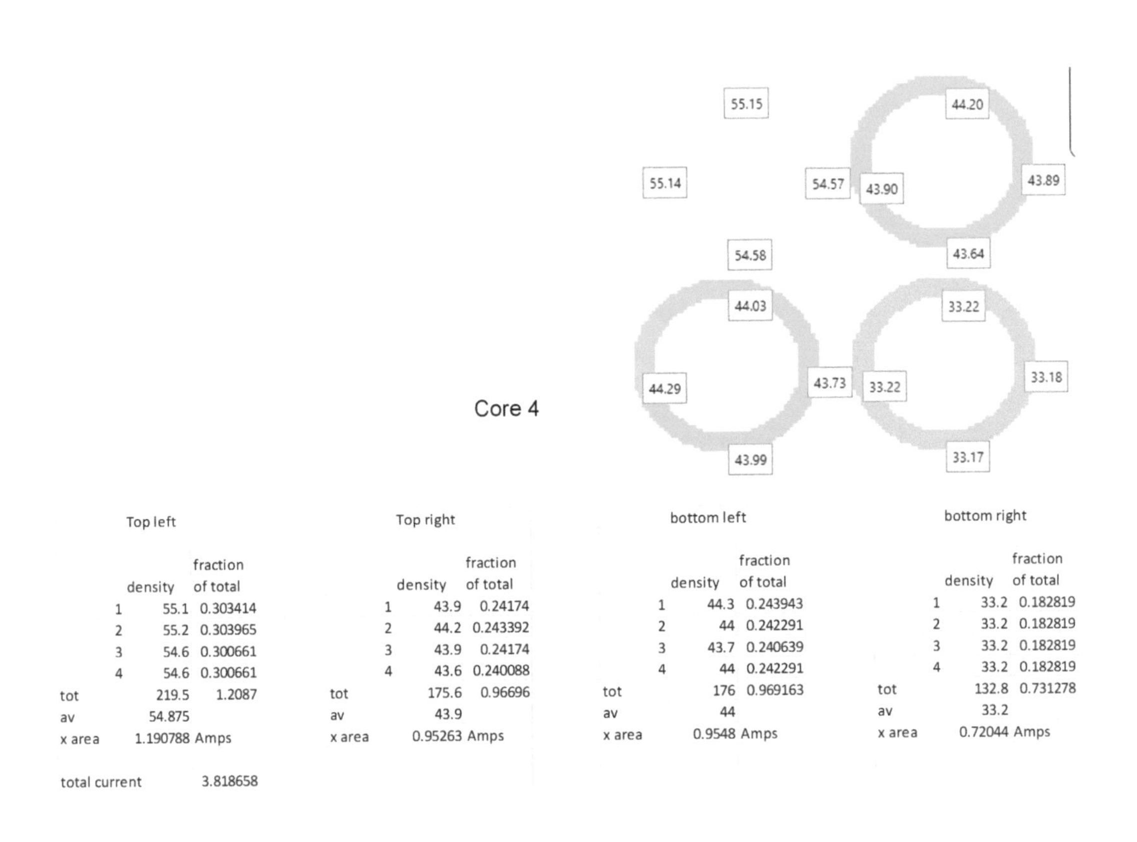

Top left

	density	fraction of total
1	55.1	0.303414
2	55.2	0.303965
3	54.6	0.300661
4	54.6	0.300661
tot	219.5	1.2087
av	54.875	
x area	1.190788	Amps

Top right

	density	fraction of total
1	43.9	0.24174
2	44.2	0.243392
3	43.9	0.24174
4	43.6	0.240088
tot	175.6	0.96696
av	43.9	
x area	0.95263	Amps

bottom left

	density	fraction of total
1	44.3	0.243943
2	44	0.242291
3	43.7	0.240639
4	44	0.242291
tot	176	0.969163
av	44	
x area	0.9548	Amps

bottom right

	density	fraction of total
1	33.2	0.182819
2	33.2	0.182819
3	33.2	0.182819
4	33.2	0.182819
tot	132.8	0.731278
av	33.2	
x area	0.72044	Amps

total current 3.818658

These figures illustrate what we mean by "rotational" symmetry. The horizontal figure (left) is the top trace layer and the vertical (right) figure is the bottom trace layer.

A7 FUSING CURRENT REFERENCE

Nomograph from Electrical World, Reference [4] in Chapter 9

SHORT-TIME CURRENT REQUIRED TO MELT COPPER CONDUCTORS*

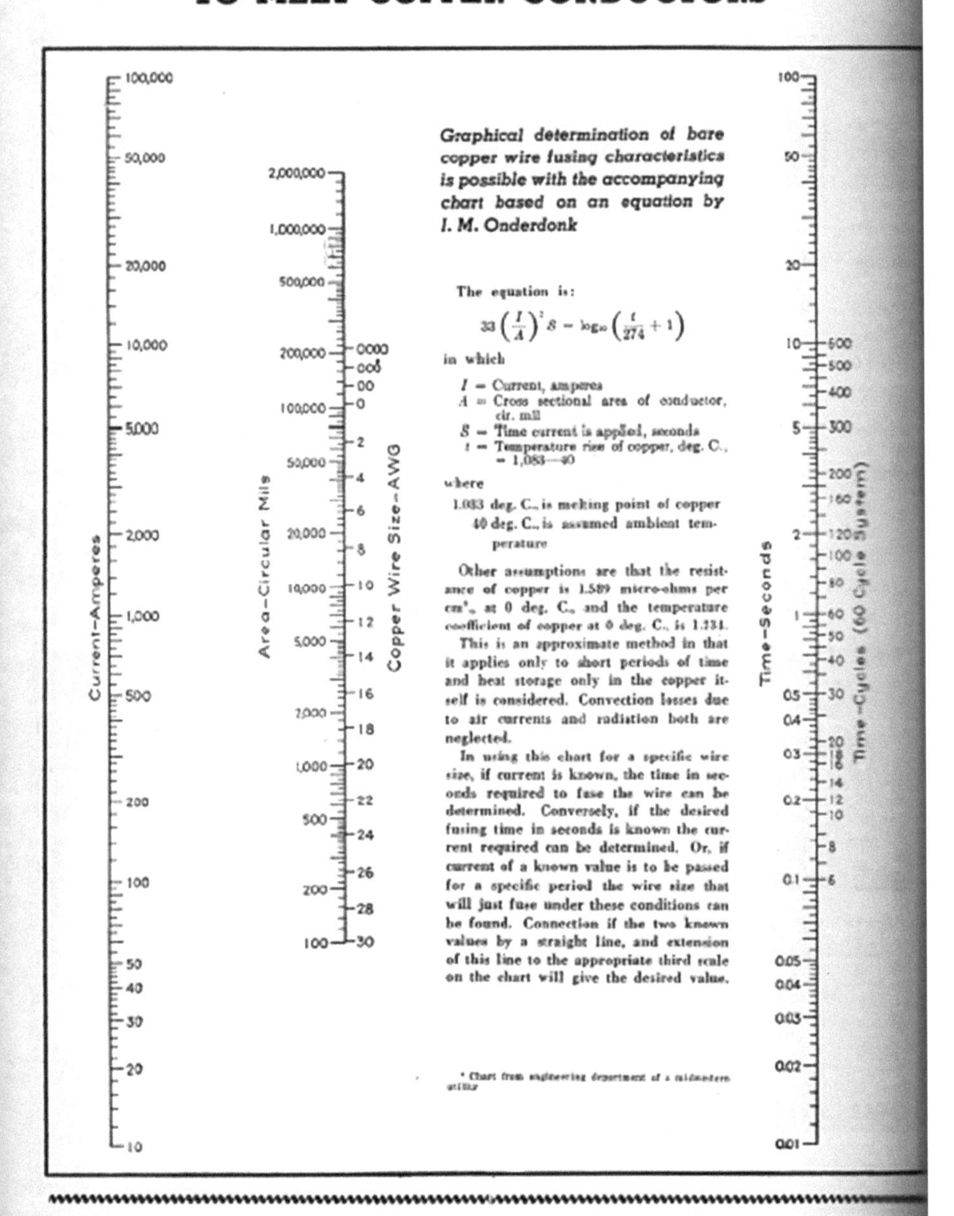

Graphical determination of bare copper wire fusing characteristics is possible with the accompanying chart based on an equation by I. M. Onderdonk

The equation is:

$$33\left(\frac{I}{A}\right)^2 S = \log_{10}\left(\frac{t}{274}+1\right)$$

in which

I = Current, amperes
A = Cross sectional area of conductor, cir. mil
S = Time current is applied, seconds
t = Temperature rise of copper, deg. C., = 1,083—40

where

1,083 deg. C., is melting point of copper
40 deg. C., is assumed ambient temperature

Other assumptions are that the resistance of copper is 1.589 micro-ohms per cm³., at 0 deg. C., and the temperature coefficient of copper at 0 deg. C., is 1.234.

This is an approximate method in that it applies only to short periods of time and heat storage only in the copper itself is considered. Convection losses due to air currents and radiation both are neglected.

In using this chart for a specific wire size, if current is known, the time in seconds required to fuse the wire can be determined. Conversely, if the desired fusing time in seconds is known the current required can be determined. Or, if current of a known value is to be passed for a specific period the wire size that will just fuse under these conditions can be found. Connection if the two known values by a straight line, and extension of this line to the appropriate third scale on the chart will give the desired value.

* Chart from engineering department of a midwestern [illegible]

A8 FUSING CURRENT SIMULATIONS

1 Oz. 20 Mil Wide Trace

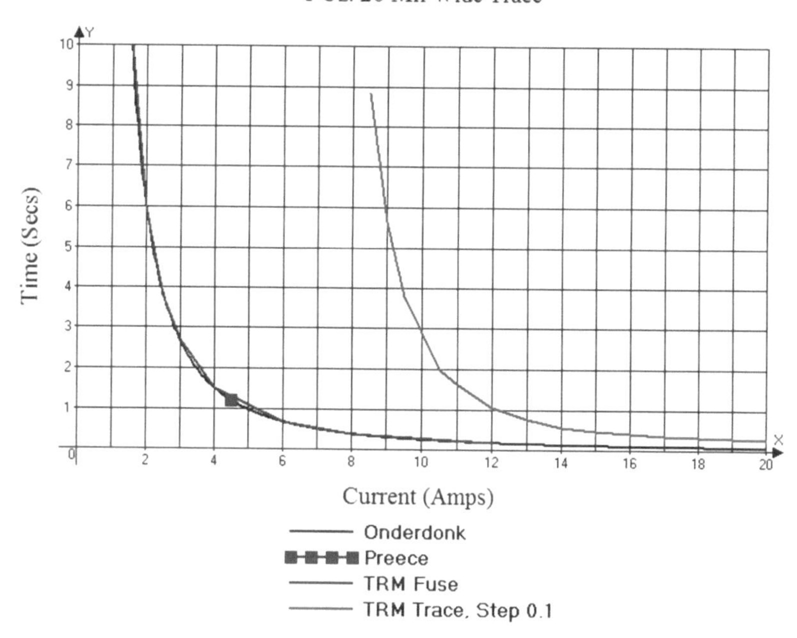

1 Oz. 100 Mil Wide Trace

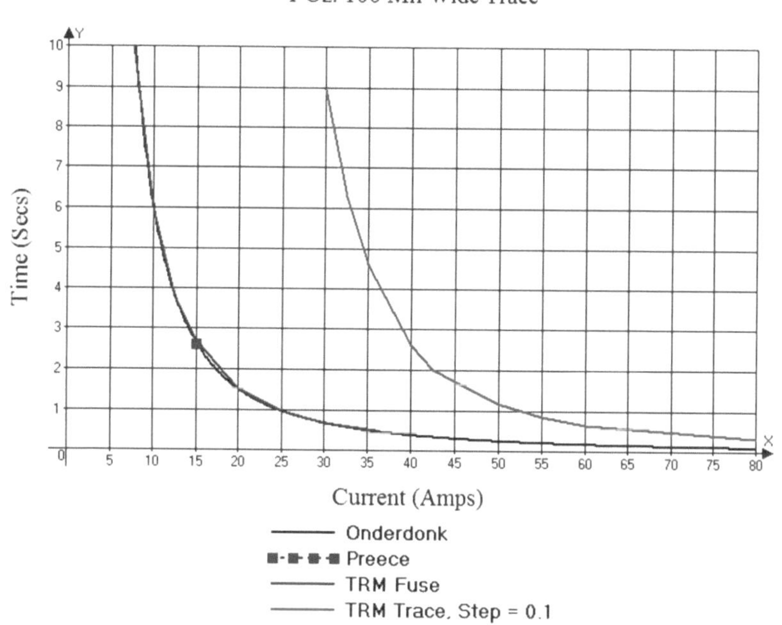

1 Oz. 200 Mil Wide Trace

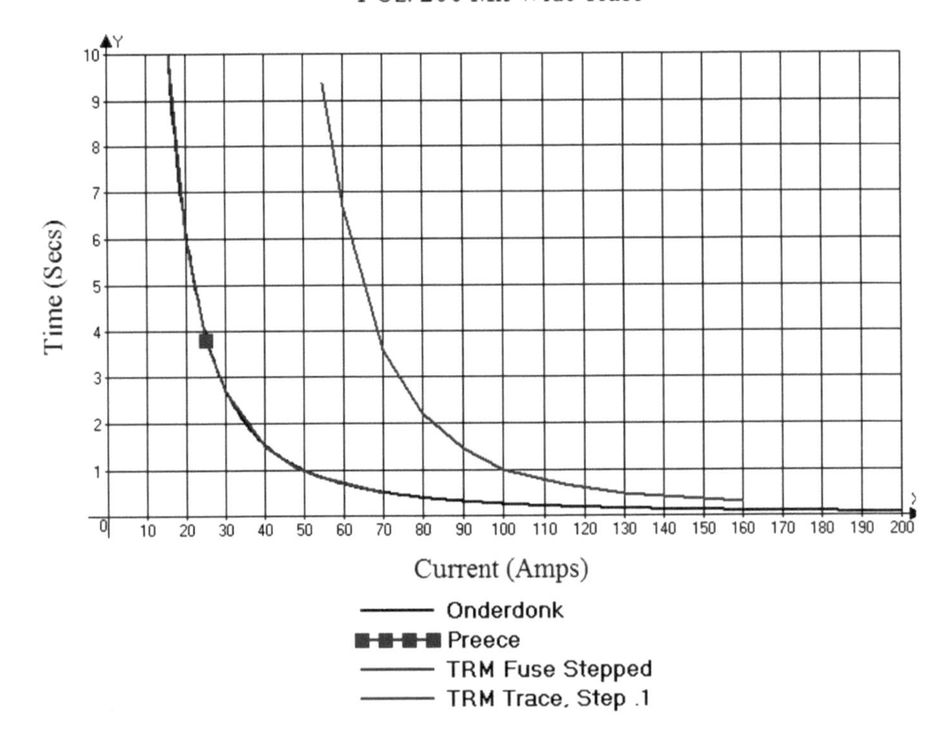

2 Oz. 20 Mil Wide Trace

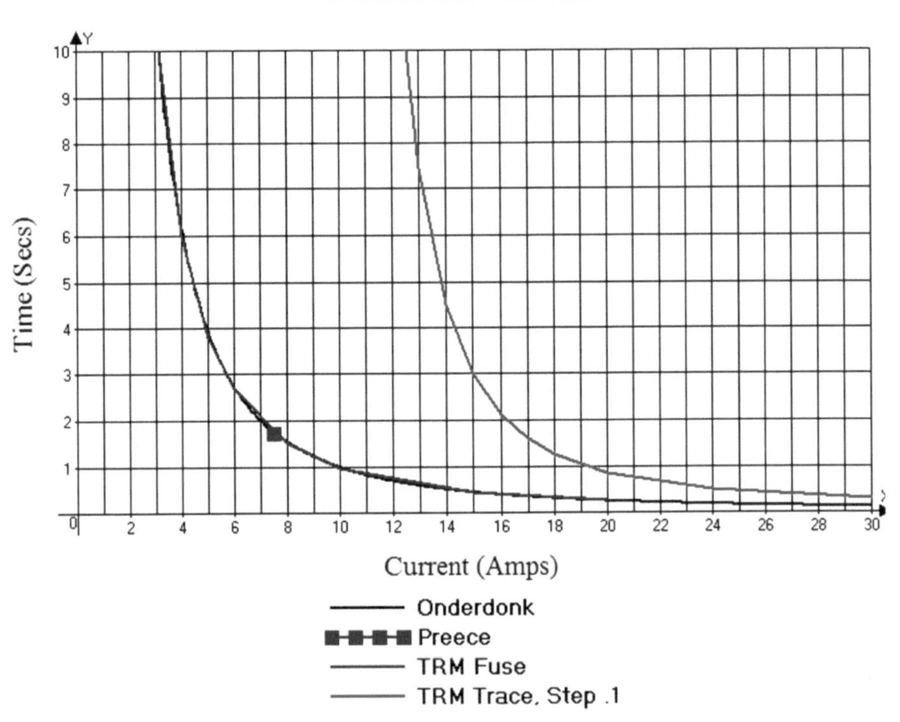

2 Oz. 100 Mil Wide Trace

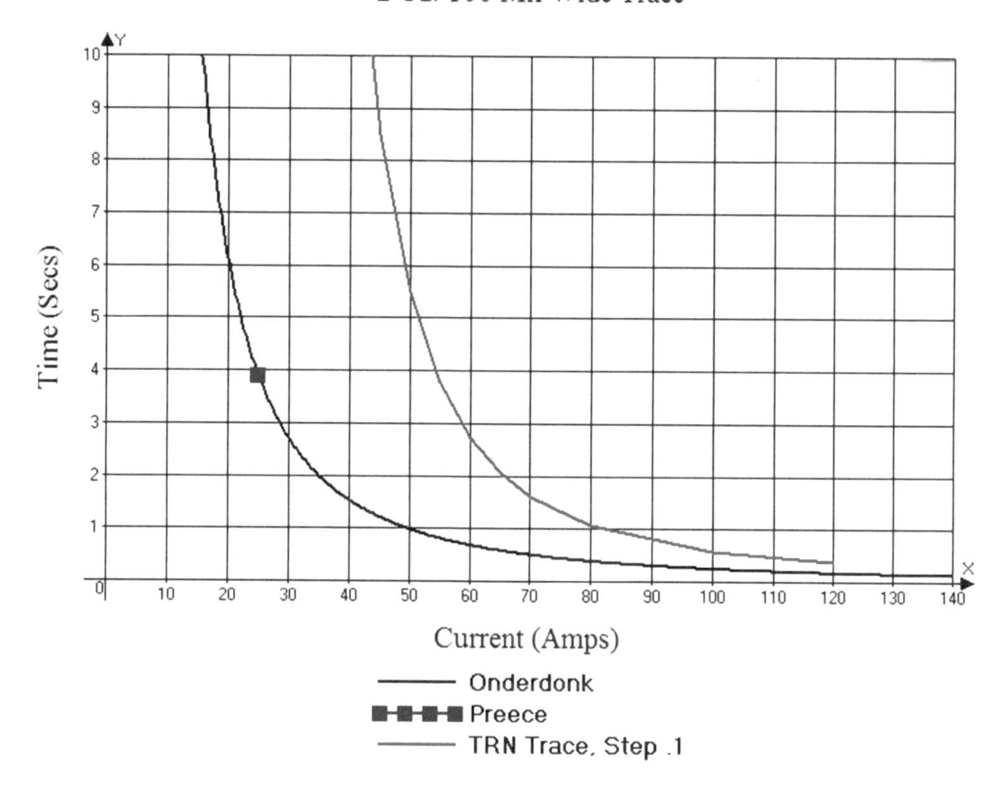

2 Oz. 200 Mil Wide Trace

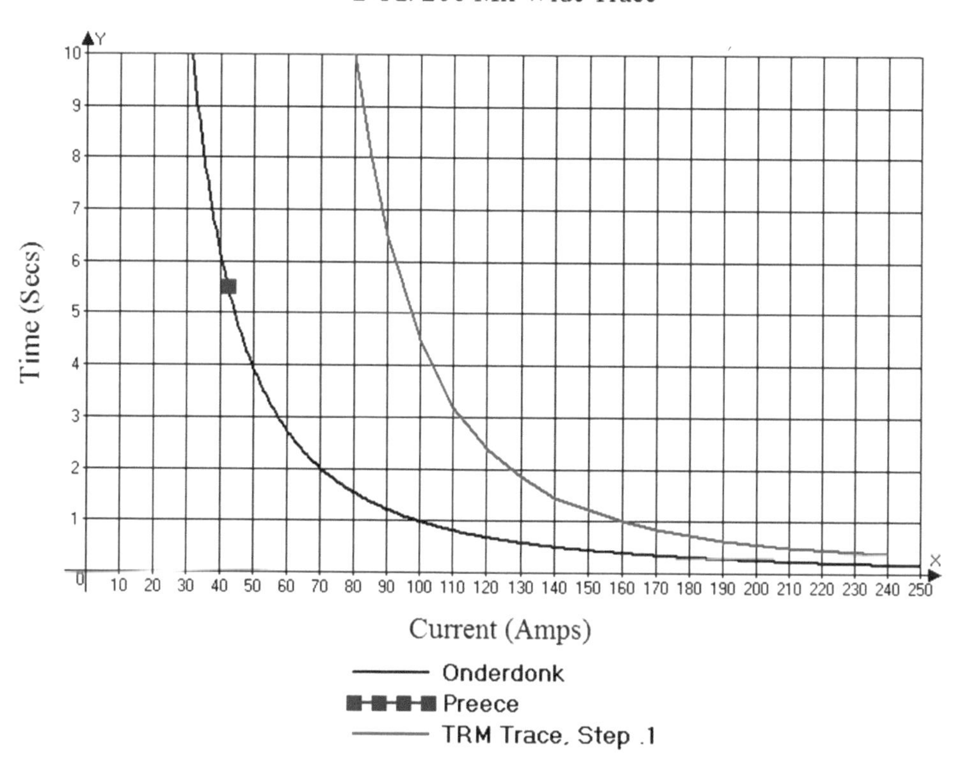

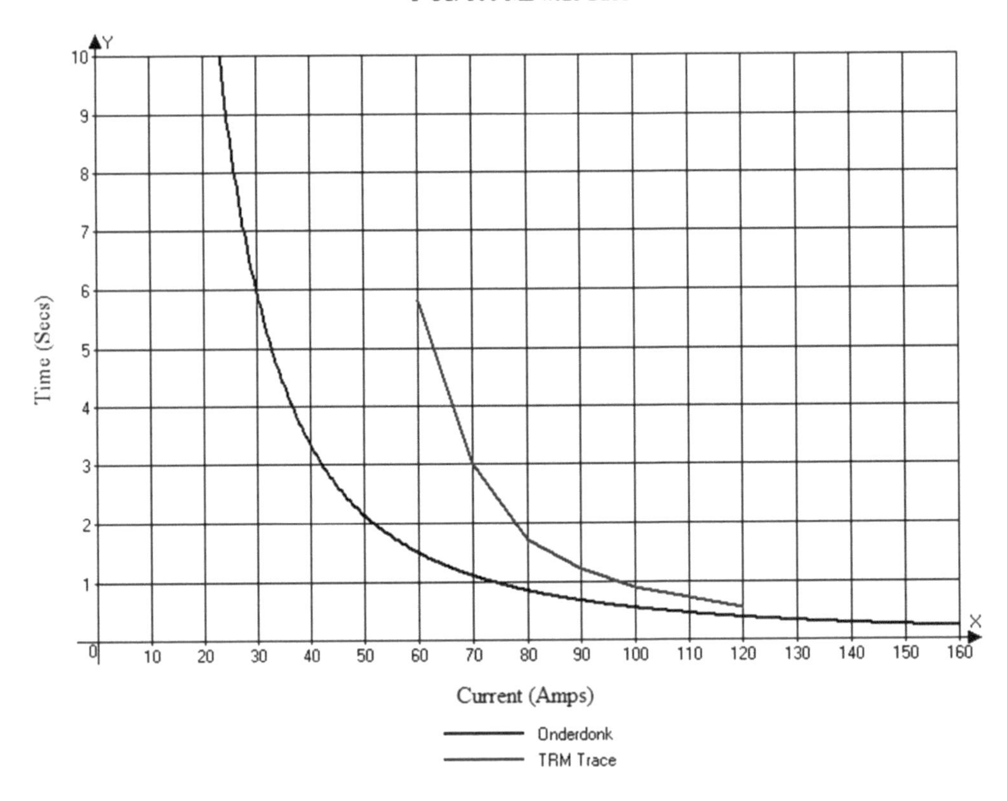
3 Oz. 100 Mll wide trace
Y
10
9
8
7
6
5
4
3
2
1
0
Time (Secs)
10
20
30
40
50
60
70
80
90
100
110
120
130
140
150
160
X
Current (Amps)
Onderdonk
TRM Trace

A9 NON-UNIFORM HEATING PATTERNS

The following images are close-ups of the thermal distribution on selected traces. The traces carried enough current to heat the trace to approximately 55° C, except where the trace was large enough that 10 Amps could not reach that temperature (10 Amps was the maximum output of the available current generator.) The images cover a very narrow temperature range, indicated by the scale under the image.

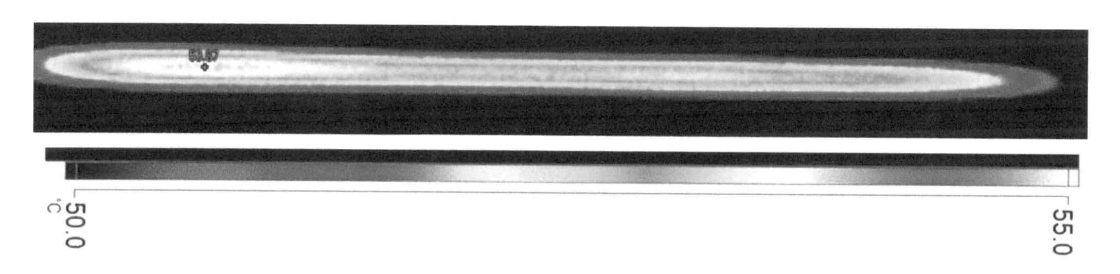

100 mil wide, 0.5 Oz. foil trace.

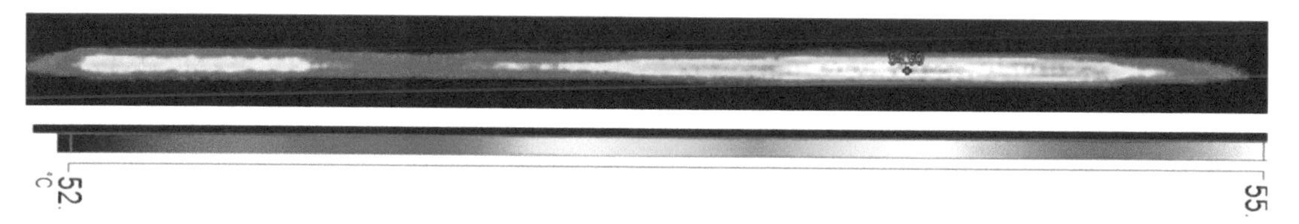

100 mil wide, 1.5 Oz. foil trace.

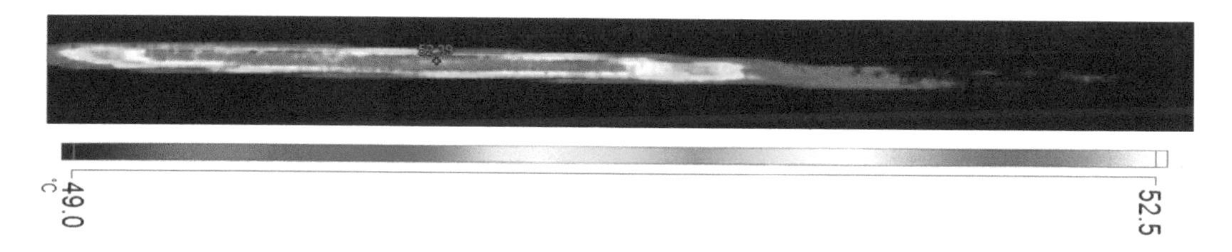

100 mil wide, 0.5 Oz. foil plated with 1.0 Oz. copper trace.

200 mil wide, 2.0 Oz. foil trace.

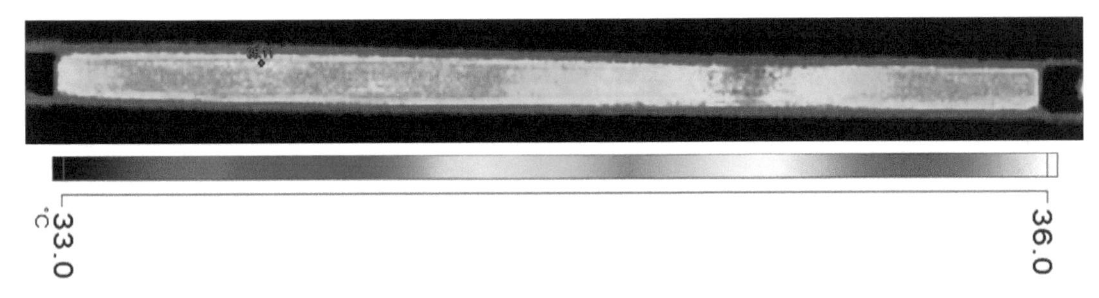

200 mil wide, 2.0 Oz. foil plated with 1.0 Oz. copper trace.

100 mil wide, 1.5 Oz. foil plated with 1.75 Oz. copper trace.

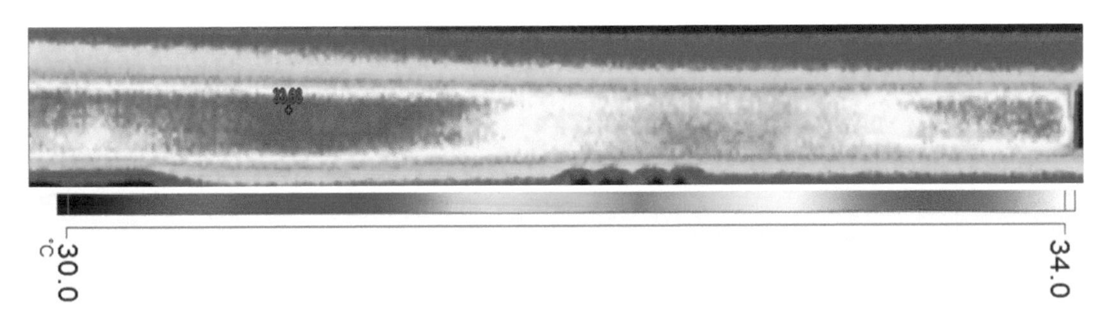

200 mil wide, 2.0 Oz. foil plated with 2.0 Oz. copper trace.

I1 ABRIDGED INDEX

Abridged index. Information about the selected items may continue past the indicated page.

absolute zero temperature 22
Ac currents 145
atom 19
Codreanu, Norocel 142, 144, 164
cooling,
    conduction 30, 130
cooling,
    HTC, see Heat Transfer Coefficient
copper,
    electrodeposited (ED) 12
copper,
    rolled 12
current density 33, 77, 162
current density, via 97
current, definition 20
current, fusing 105
current, fusing, time, 120
    slight overload 131
    strong overload 131
current, laws of 26
dielectric properties,
    decomposition temperature 17, 131
    glass transition temperature 16, 131
    thermal conductivity 16, 30
        in-plane 31, 90, 177
        through plane 31, 90, 177
    time to delamination (T260/T288) 17, 131
electron velocity 26
heating, i-squared R 29, 129
hot spots 130, 139
HTC, heat transfer coefficient 31, 54, 130
I^2t (constant) 124, 128
impedance controlled traces 9
IPC 2152 8, 11, 43
    external 45
    external, equations 46
    internal, equations 49
    vacuum, equations 50

IPC 2221 8
IPC-D-275 8
Matula, R. A. 23
microsection 25
Mil Std 1495 7
Mil std 275E 8
model, trace heating and cooling 33
NBS 4283 7
Onderdonk equation 107
    derivation 110
Onderdonk, I. M. 107
panel plating 13
pattern plating 13
plating thickness 14
power distribution network (PDN) 4
power/energy 30
Preece, W. H. 105
resistance 21
resistivity 15
resistivity 19
    liquid copper 24
    measuring 24, 179
    table, copper 24
    thermal coefficient of 22
right-angle corners 159
RMS current 145, 154
sizing traces 3
Stauffacher, E. R. 108
temperature, trace, measuring 34
    change of resistance 34
    infrared 35
    thermocouple 36
tg, see dielectric properties
thermal curve, typical 37
thermal runaway 39,134
thermograph 91, 217
TRM (Thermal Risk Management) 8, 53
TRM modeling process 56
TRM sensitivity analyses,
    adjacent trace 67
    air flow 69
    presence of planes 66
    small trace width 62
    thermal gradient 64
    trace and plane 68
    trace length 64
    transient response 65
via temperature 81
via temperature, cooler than trace 84, 90, 94
via temperature, two vias 88
via, thermal 91
Voltage drops 93
x-ray 26, 167

ABOUT THE AUTHORS

Douglas Brooks has a BS/EE and an MS/EE from Stanford and a PhD from the University of Washington. For the last 20 years he has owned a small engineering service firm and written numerous technical articles on Printed Circuit Board Design and Signal Integrity issues, and has published two books. He has given seminars several times a year all over the US, as well as Moscow, China, Taiwan, Japan, and Canada. His primary focus is on making complex technical issues easily understood by those without advanced degrees.

Johannes Adam got a doctorate in physics from University of Heidelberg, Germany, in 1989 on a thesis about numerical treatment of 3- dimensional radiation transport in moving astrophysical plasmas. He was then employed in software companies, mainly working on numerical simulations of electronics cooling at companies like Cisi Ingenierie S.A. , Flomerics. Ltd. and Mentor Graphics Corp. In 2009 he founded ADAM Research and does work as a technical consultant for electronics developing companies and as a software developer. He is the author of a simulation program called TRM (Thermal Risk Management), designed for electronics developers and PCB designers who want to solve electro-thermal problems at the board level. He is member of the German chapter of IPC (FED e.V.) and engages in its seminars about thermal topics. He is Certified Interconnect Designer (CID). He is living in Leimen near Heidelberg.

CPSIA information can be obtained
at www.ICGtesting.com
Printed in the USA
BVHW021959210119
538254BV00031B/177/P

* 9 7 8 1 5 4 1 2 1 3 5 2 4 *